AF477886

ADVANCED VERIFICATION TECHNIQUES:
A SystemC Based Approach for Successful Tapeout

ADVANCED VERIFICATION TECHNIQUES:
A SystemC Based Approach for Successful Tapeout

by

Leena Singh
Azanda Network Devices
Leonard Drucker
Cadence Design Systems
Neyaz Khan
Cadence Design Systems Inc.

KLUWER ACADEMIC PUBLISHERS
Boston / Dordrecht / New York / London

Distributors for North, Central and South America:
Kluwer Academic Publishers
101 Philip Drive
Assinippi Park
Norwell, Massachusetts 02061 USA
Telephone (781) 871-6600
Fax (781) 871-6528
E-Mail <kluwer@wkap.com>

Distributors for all other countries:
Kluwer Academic Publishers Group
Post Office Box 322
3300 AH Dordrecht, THE NETHERLANDS
Telephone 31 78 6576 000
Fax 31 78 6576 474
E-Mail <orderdept@wkap.nl>

Electronic Services <http://www.wkap.nl>

Library of Congress Cataloging-in-Publication

Title: Advanced Verification Techniques:
A SystemC Based Approach for Successful Tapeout
Author (s): Leena Singh, Leonard Drucker
Neyaz Khan
ISBN: 1-4020-7672-X
ISBN: 1-4020-8029-8 (eBook)

Contents

Authors xi
Acknowledgements xiii
Foreword xv

CHAPTER 1   *Introduction* 1

1.1 Verification Overview and Challenges 3

     1.1.1 Challenges 5

1.2 Topics Covered in the book 7

     1.2.1 Introduction 7

     1.2.2 Verification Process 7

     1.2.3 Using SCV for Verification 7

     1.2.4 Test Plan 7

     1.2.5 Test Bench Concepts using SystemC 8

     1.2.6 Methodology 8

     1.2.7 Regression 8

     1.2.8 Functional Coverage 9

     1.2.9 Dynamic Memory Modeling 9

     1.2.10 Post Synthesis/Gate Level Simulation 9

1.3 Reference Design 10

     1.3.1 Networking Traffic Management and SAR chip ... 10

     1.3.2 Multimedia Reference Design 14

CHAPTER 2   *Verification Process* 17

2.1 Introduction 18

2.2 Lint 21

2.3 High Level verification languages 22

2.4 Documentation 26

     2.4.1 Documenting verification infrastructure 27

2.5 Scripting 30

2.6 Revision control 32

2.7 Build Process 35

2.8 Simulation and waveform analysis 35

     2.8.1 Waveform Analysis tool features 36

2.9 Bug tracking 37

2.10 Memory modelling 39

2.11 Regression 40

2.12 Functional Coverage 40

     2.12.1 Requirements for a Functional Coverage Tools ... 40

     2.12.2 Limitations of Functional Coverage 41

	2.13 Code Coverage	41
CHAPTER 3	*Using SCV for Verification*	45
	3.1 Features a verification language	45
	3.2 Why use C++ for SCV?	47
	3.3 What is randomization	48
	3.4 Introduction to SCV	49
	3.4.1 Basic Purpose of SCV	49
	3.4.2 Basic Language Constructs of SCV	50
	3.4.3 Examples of an SCV testbench	53
	3.5 Creating A Stimulus Generator using SCV	54
	3.5.1 Testbench Structure	54
	3.5.2 Randomization	56
	3.6 Future of SCV	64
CHAPTER 4	*Functional Verification Testplan*	65
	4.1 Different kinds of Tests	66
	4.2 When to Start	68
	4.3 Verification plan document	70
	4.4 Purpose and Overview	70
	4.4.1 Intended Use	70
	4.4.2 Project References	70
	4.4.3 Goals	70
	4.4.4 System Level Testbench:	72
	4.4.5 DUV	73
	4.4.6 Transaction Verification Modules	74
	4.4.7 Packet/Cell Generators	75
	4.4.8 CPU Command Generators	76
	4.4.9 Packet/Cell Checkers	76
	4.4.10 System Level Checker	77
	4.4.11 Rate Monitors	77
	4.4.12 Simulation Infrastructure	78
	4.4.13 Verification Infrastructure	80
	4.4.14 System Testing	83
	4.4.15 Basic Sanity Testing	83
	4.4.16 Memory and Register Diagnostics	83
	4.4.17 Application types sanity tests	84
	4.4.18 Data path Verification	84
	4.4.19 Chip Capacity Tests	85
	4.4.20 Framer Model	86
	4.4.21 Congestion Check	86
	4.4.22 Customer Applications	86
	4.4.23 Free Buffer List checking	86

4.4.24 Overflow FIFOs 86
4.4.25 Backpressure 86
4.4.26 Negative Tests 86
4.4.27 Block Interface Testing 87
4.4.28 Chip feature testing 87
4.4.29 Random Verification 89
4.4.30 Corner Case Verification 90
4.4.31 Protocol Compliance checking 90
4.4.32 Testing not done in simulation 90
4.4.33 Test Execution Strategy 90
4.5 Tests Status/Description document 91
4.6 Summary 93
CHAPTER 5   *Testbench Concepts using SystemC* 95
5.1 Introduction 95
5.1.1 Modeling Methods 95
5.1.2 Transaction Level Modeling 96
5.1.3 Block level Verification using TLM 97
5.2 Unified Verification Methodology(UVM). 103
5.2.1 Functional Virtual Prototype 103
5.2.2 Transactions 107
5.2.3 Assertions 109
5.2.4 Considerations in creating an FVP 110
5.2.5 Creating the FVP 113
5.3 Testbench Components 117
5.3.1 Verification Communication Modes 119
5.3.2 Using Transactions in testbenches 123
5.3.3 Characteristics of Transactions 124
5.3.4 Hierarchical Transactions 125
5.3.5 Related Transactions in multiple streams 125
CHAPTER 6   *Verification Methodology* 149
6.1 Introduction 150
6.2 Overall Verification approach 150
6.3 What all tools are needed 152
6.4 Transaction based verification environment: 153
6.4.1 Elements of TBV 154
6.5 Design verification 155
6.5.1 System Level Verification 156
6.5.2 Block Level Verification 159
6.5.3 Block TVMs at system level: Interface TVMs: . . . 160
6.6 Verification infrastructure 161
6.7 Interface TVMs 162

6.7.1 PL3 163
6.7.2 Other interfaces 170
6.8 Traffic Generation 170
6.8.1 Traffic generator methodology 170
6.8.2 Overall Flow 172
6.8.3 Implementation flow 174
6.8.4 Interface Driver hookup to traffic Generation 175
6.8.5 To setup test stimulus for generation: 176
6.9 Writing Stimulus File 177
6.9.1 Example stimulus file front end 177
6.10 Monitors 193
6.11 Checkers 193
6.11.1 Methodology used in reference design 194
6.12 Message Responder 201
6.12.1 Flow 201
6.12.2 Basic Structure 203
6.12.3 Example code 203
6.13 Memory models 209
6.14 Top Level simulation environment 209
6.15 Results of using well defined methodology 210

CHAPTER 7   *Regression/Setup and Run*   211
7.1 Goals of Regression 212
7.2 Regression Flow and Phases 213
7.2.1 Regression Phases 214
7.3 Regression Components 216
7.3.1 Bug tracking 216
7.3.2 Hardware 217
7.3.3 Host Resources 217
7.3.4 Load Sharing Software 218
7.3.5 Regression Scripts 219
7.3.6 Regression Test Generation 219
7.3.7 Version control 220
7.4 Regression features 220
7.4.1 Important features for regression 220
7.4.2 Common switches for Regression 224
7.5 Reporting Mechanism 225
7.5.1 Pass/fail count Report 226
7.5.2 Summary for each test case 227
7.5.3 Verbose Report for Debugging if error occurs . . . 227
7.5.4 Error diagnosis 227
7.5.5 Exit criteria 227

7.5.6 Verification Metrics 228
7.6 Common Dilemmas in Regression 228
7.7 Example of Regression run 229
7.7.1 Run options. 229
7.8 Summary 231

CHAPTER 8   *Functional Coverage* 233
8.1 Use of Functional Coverage. 234
8.1.1 Basic Definition Functional Coverage. 234
8.1.2 Why Use a Functional Coverage? 234
8.2 Using Functional Coverage in Verification environment. 236
8.2.1 Functional and Code Coverage Difference. 236
8.3 Implementation and Examples of Functional Coverage. .237
8.3.1 Coverage Model Design. 238
8.3.2 Transaction Functional Coverage techniques. 241
8.3.3 Functional Coverage Examples 243
8.3.4 Role of Functional and Code coverage 245
8.4 Functional Coverage Tools. 245
8.4.1 Commercial Functional Coverage Tools. 245
8.4.2 Features of a good functional coverage tool. 246
8.4.3 Requirements for a Functional Coverage Tools . . .248
8.5 Limitations of Functional Coverage 250

CHAPTER 9   *Dynamic Memory Modeling* 251
9.1 Various solutions for simulation memory models 252
9.2 Running simulation with memory models 252
9.2.1 Built in memory models .252
9.3 Buying commercially available solutions 255
9.4 Comparing Built in and Commercial memory models . . .256
9.5 Dynamic Memory modeling Techniques 260
9.5.1 Verilog 261
9.5.2 Using Wrapper 262
9.6 Example From Reference Design: 264
9.6.1 Performance comparison 273

CHAPTER 10   *Post Synthesis Gate Simulation* 275
10.1 Introduction 276
10.1.1 Need for Gate Level Simulation 276
10.2 Different models for running gate netlist simulation . . .277
10.2.1 Gate Simulation with Unit Delay timing 277
10.2.2 Gate-level simulation with full timing 277
10.3 Different types of Simulation .278
10.3.1 Stages for simulation in design flow 280
10.4 Getting ready for gate simulation 281

10.4.1 Stimulus generation 281
10.4.2 Back annotating SDF files 282
10.4.3 Multicycle path 282
10.4.4 Using Sync Flops 283
10.4.5 ncpulse and transport delays 284
10.4.6 Pulse handling by simulator 285
10.5 Setting up the gate level simulation 287
10.6 ATE vector Generation 289
10.6.1 Requirement for Functional testing on ATE 289
10.6.2 Various ATE Stages 290
10.6.3 Choosing functional simulations for ATE 290
10.6.4 Generating vectors 291
10.6.5 Applying vectors back to netlist 291
10.6.6 Testing at ATE 291
10.6.7 Problem at speed vectors 292
10.7 Setting up simulation for ATE vector generation 293
10.8 Examples of issues uncovered by gate simulations 294
APPENDIX   297
0.1 Common Infrastructure 297
0.2 Simple example based on above methodology 299
0.2.1 Generator 299
0.2.2 Sideband 303
0.2.3 Driver 304
0.2.4 Monitor 310
0.2.5 Checker 314
0.2.6 Stim 318
0.2.7 SimpleRasTest.h 318
0.2.8 SimpleRasTest.cc 318
0.3 Example for standard interfaces: 319
0.3.1 Code Example for PL3Tx Standard interface: . . . 319
0.3.2 Code Example for Standard PL3Rx interface 332
0.3.3 SPI4 interface 338
0.3.4 Code Example for SPI4 interface 344
References 371
Index. 373

Authors

Leena Singh has over 12 years experience in ASIC design and verification for networking, multimedia, wireless, and process control applications. She led various teams responsible for delivering methodologies for SOC verification using advanced verification techniques. She was also in charge of the verification effort at Networking and Multimedia startups through entire cycle of setting up the verification team to tapeouts. Leena graduated with a BSEE from Punjab University, Chandigarh, in India. She is currently Senior Member of Consulting Staff in the Methodology Engineering group of Cadence Design Systems, Inc.

Leonard Drucker is currenlty working as an Architect in Core Competency group at Cadence Design Systemc, Inc. He is instrumental in product definition and strategy, helping shape and define verification technologies with product management, marketing and sales. Prior to Cadence, he served as an independent consultant, focusing on digital verification. His extensive design verification and simulation background also includes key positions at EDA companies Zycad, Solbourne, Cadnetix and Medtronic. Drucker earned his Bachelor of Science degree in electrical engineering from Arizona State University.

Neyaz is currently working as an Architect in Core Competency group at Cadence Designs systemc, Inc. Neyaz has over 15 yrs of experience on all aspects of front-end ASIC design from transistor-level design of CMOS elements to complex DSP based architectures of modern wireless communications and broadband networking systems. Prior to working for Cadence Neyaz has served in key technical-lead roles

---

in the Design and Verification of complex, multi-million gate ASICs for wireless and telecom applications for a number of companies including Texas Instruments and Bell Northern Research (now Nortel Networks). Neyaz earned his Bachelor of Engineering degree in Electronics and Telecom from REC Srinagar in India, and a Masters of Applied Sciences degree in Electrical Engineering from Concordia University, Montreal, Canada.

Acknowledgements

Writing a book is a difficult and time consuming endeavor. It requires support from many different people to make it happen. Authors would like to thank Paul Wilcox, Andreas Meyers, John Rose, John Pierce, technical leaders at Cadence, for their technical contributions and collaborations for this book. Authors would also like to thank their management, Jose Fernandez and Chris Dietrich, for allowing them to pursue this activity.

Special thanks to the verification team at Azanda Networks: Shawn Ledbetter, Brahmananda Marathe, Mathew Springer, Rajesh Kamath, Sophia Arnaudov for their valuable contributions to build the verification environment that we are all proud of. It was great experience to show the value of using right methodology in meeting aggressive startup schedules.

Special thanks to Shankar Channabasan, Design Manager at Azanda Networks, for reviewing the chapters and providing useful information in Post Synthesis/Gate Simulation chapter.

Special thanks to Heath Chambers, President/Verification Designer, HMC Design Verification, Inc. for reviewing the book and providing his valuable suggestions.

In addition, each author would like to add their individual acknowledgements.

Leena Singh: I would like to express my gratitude to my father, Surinder Bedi, husband, Sarabjeet Singh, daughter, Keerat, and son, Jaspreet for their love, patience, and unconditional support while writing this book.

Leonard Drucker: First, I would like to thank my wife Ivonne and my children, Kayle, Jasmine, and Blake for allowing me to focus on writing this book when I should have been participating in family activities. Next, I'd like to thank my parents, Harold and Florence Drucker, for their instilling in me an ambitious drive and a strong belief in myself.

Neyaz Khan: First of all, I appreciate and acknowledge the contributions made by my beautiful wife Shamila to this book. Without her, life is not only unmanageable, but impossible. Thanks for letting me spend those hours in front of the computer. Next, I would like to acknowledge my three lovely children Aimun, Ayan, and Anum, who are a constant source of inspiration and never fail to bring a smile to my face. Last, but not the least my parents, Tarique & Nigar, for raising me the right way, and giving me the courage and confidence to believe in myself.

Verification is an extremely important task in any design flow and is a requirement for every design that is taped out. This book makes an attempt to address the practical needs of the design and verification community to take a step forward towards uniform verification methodology. It will be extremely satisfying to the authors if the contents of the book are found useful by the readers.

Leena Singh

Leonard Drucker

Neyaz Khan

San Jose, California

April, 2004

The last few years in the electronics industry have seen three areas of explosive growth in design verification. The first 'explosion' has been the tremendous increase in verification requirements for IC's, ASICs, ASSPs, System-on-Chip (SoCs) and larger systems. This has been compounded by the rapid increase in use of embedded processors in these systems from the largest network infrastructure boxes all the way down to consumer appliances such as digital cameras and wireless handsets. With processors deeply embedded in systems, verification can no longer remain solely focused on hardware, but must broaden its scope to include hardware-dependent software co-verification, and large components of middleware and applications software verification in the overall system context.

The second area of explosive growth has been in the number and variety of verification methodologies and languages. New concepts including transaction-based verification (TBV), transaction-level modelling (TLM), and assertion-based verification (ABV) have arisen to complement the already-extensive verification engineer's toolbox. These new verification approaches are not just theoretical - they have been growing in their application to solve practical and vital verification problems. Complementing the new verification approaches has been a rapid growth in verification languages - first, specialized Hardware Verification Languages (HVLs) such as Verisity's 'e' (now being standardized by the IEEE 1647 working group); Synopsys Vera (now subsumed in SystemVerilog) and Cadence Testbuilder (a C++

extension class library and methodology, now largely subsumed in the SystemC verification library (SCV)). The HVLs as stand-alone notations and languages provided substantial increases in verification quality and functional coverage, and have led rather naturally to the idea of an integrated Hardware Design and Verification Language (HDVL). These include pioneering experiments such as Co-Design Automation's SUPERLOG (now subsumed in SystemVerilog), Accellera's development of SystemVerilog (to be donated to the IEEE 1364 committee in summer 2004 for standardization as part of Verilog evolution), and Open SystemC Initiative's (OSCI) SystemC. SystemC is a particularly interesting example of an HDVL - starting off with a Register-Transfer Level (RTL) design language in C++ with SystemC 1.0, extending the language to generalized system modelling with SystemC 2.0, and adding the SystemC Verification Library (SCV) in 2003 to form a full "SMDVL - System Modelling, Design, and Verification Language". Compounding the alphabet soup of HDLs and HVLs are the assertion and property libraries and languages - Open Verification Library (OVL), OpenVera Assertions (OVA), and Accellera's Property Specification Language (PSL) and SystemVerilog Assertions (SVA).

The third explosion in verification is one that is just starting - the growing availability of books full of guidance and advice on verification techniques, verification languages, verification tools, and ways of bringing them all together into a coherent verification methodology and process. This new volume, *Advanced Verification Techniques, A SystemC Based Approach for Successful Tapeout,* by Leena Singh, Leonard Drucker and Neyaz Khan, is unique among all the books on verification that are now exploding into existence. As such, it belongs on the shelf of every design and verification engineer - but, much more than being "on the shelf", it belongs on the desk of every engineer, preferably opened to a relevant page of verification guidance or knowledge which is being pressed into active service.

What makes this book especially important and unique is that it is the first verification book which gives detailed insights into C/C++ based verification methods, by focusing on the usage of TestBuilder and the SystemC Verification library (SCV). In fact, the book by Singh, Drucker, and Khan has some of the earliest examples in print of SCV used in action. It covers all the major processes involved in a modern functional verification methodology, and thus complements earlier books concentrating on verification tools, and design and verification languages.

SystemC and C++ are particularly good bases on which to build a verification approach. The use of C++ allows linkages to be built between all participants in the design and verification process. System-level architects can build early algorithmic and architectural models at high levels and transaction levels which are useful both for establishing system level performance characteristics, and for acting as golden models during downstream verification processes. The scenarios which stress the system at a high level can become corner cases for functional correctness. The use of C++ as a base allows for easier reuse of testbenches and models between all participants in the process. C++ allows easy integration of instruction set simulators, so hardware-software co-verification becomes an attribute of the verification process. Data type abstraction, templating and other abstraction features offered by C++, SystemC, Testbuilder and SCV all contribute to verification productivity, reuse, coverage and quality.

Advanced Verification Tecniques, A SystemC Based Approach for Sucessful Tapeout, after an introduction to verification and the overall book, discusses the verification process in chapter 2, and then follows that with an extensive introduction to the use of SCV. Chapter 4 discusses the development of the functional verification testplan, which consolidates the goals, methods and detailed implementation of the verification requirements into a comprehensive plan for verifying a design. Chapter 5 describes in detal the basic concepts of modern testbenches, and links them to SystemC. This includes detailed guidance on linking SystemC testbenches to RTL designs using HDLs. Chapter 6 gives a good overview of transaction based verification methodology and explicit examples of putting this into practice in creating testbenches. These three chapters, 4-6, give a very clear message - verification of design is not an afterthought; it is a process that must be planned in advance, before designers start their work; be well-architected; have clear goals and well laid out methods to achieve them; and, deserves as much attention as the design process itself.

Special chapters, 7-10, discuss particular important topics in the verification domain. Regression, for example, is the backbone of a high-quality verification process as the design and verification tasks proceed. Regression allows the design to proceed and gives a mechanism for ensuring that design changes move in a positive direction. Regression adds tremendous confidence to a design team and its management, so that the decision to move onto detailed implementation, and the decision to release a design to manufacturing, are based on science, not just gut feelings and art.

Functional coverage complements the use of regression in ensuring adequate verification has taken place. Chapter 8 on functional coverage gives a good overview of this topic and specific guidelines and examples on how to apply it. A further chapter on memory modelling gives an introduction to an important aspect of verification that is often overlooked in these kinds of books. This is followed by a discussion, in the final chapter, on post-synthesis verification at the gate level, which also deals with functional testing and ATE (automated test equipment) - again, important pragmatic topics which are often ignored if one were focused entirely on front-end verification.

Complementing the guidelines and advice in the book are extensive examples in both Testbuilder (which can be regarded as the first or an early generation of SystemC's Verification library) and SCV, and based on two example designs - one from networking and one from multimedia. These give an excellent and pragmatic illustration of how to apply the methods, languages and tools in practice. The important thing about the examples is that they are drawn from real development work of real designs, and thus proven in anger.

To sum up, the new book from Singh, Drucker and Khan is an important milestone in the development of C/C++-based verification methods, and I recommend it highly to anyone who wishes to learn about verification within a C++-based context. We can expect further use of SystemC verification, and this book will help all those who want to apply these methods to their practical design verification problems on a daily basis.

Grant Martin
Chief Scientist
Tensilica Inc. Santa Clara

 Introduction

This book is for engineers working on ASIC/SOC design and verification as well as students or beginners who want to enhance their knowledge about how to actually verify a chip. Beginners can get insight of what is expected. For verification experts it will be a guide to help improving their methodologies for efficient and effective functional verification processes. Topics covered in this chapter are:

- Purpose of the book
- General info on importance of verification and challenges
- Brief introduction on rest of the chapters
- Reference designs

The main purpose of this book is to present the whole process of functional verification of an ASIC for successful tape out, with examples from first pass working silicon of multi million gate designs. The idea is to present a successful methodology so that a trend can be set to move towards a uniform verification methodology where everyone can share various verification components with each other. This book is a guide for verification engineers to help improve the methodologies for efficient and effective functional verification processes.

It is meant for everyone in industry involved with functional verification of ASICs. Also, it is useful for students for understanding what are the tools/processes to complete functional verification of ASICs. Design/verification engineers who are working on ASIC or SOC verification can use this book for self study. It can also be used as reference book in verification courses and any related topics.

Using the same reference designs throughout the book offer practical illustration of all the functional verification concepts for implementing reusable, independent and self-checking tests. The examples used for standard interfaces or designs are complete and not just the snippets of code and can be used as an IP. It not only covers different aspects of verification but also defines all different processes to be followed for functional verification of a chip, with clear detailed examples focusing on usage of object oriented languages.

Attention is paid to using the newer techniques for verification such as SCV, assertion based techniques, and object oriented C++ test bench. Also, using these most advanced techniques will help the industry improve the supporting tools for verification and achieve their goal for a uniform methodology where various verification teams scattered in different companies can reuse the work and other verification components.

To provide a comprehensive overview and help gain stronger understanding of the process of functionally verifying your multi million device, this book has extensive coverage of topics like –

- Verification processes
- Using SCV for verification
- Test plan
- Writing test benches
- Verification infrastructure
- Regression setup and running
- Functional Coverage
- Dynamic memory modeling
- Post synthesis simulations and verification

Topics covered in the book are not dependent on any specific tools or processes that will become out of date. It is focused on languages, tools and methodologies that verification engineers have started to adapt for future extensions for better and faster verification. This book will cover various processes followed for effective and efficient functional verification that are necessary for first pass silicon success

1.1   Verification Overview and Challenges

Verification begins at the architectural stage of the project. During this period, verification engineers are mapping out a complete development infrastructure such as database, revision control, job control mechanism etc. as well as assisting in the evaluation of various IP cores. If a development infrastructure exists then they are determining how to improve it to deal with the upcoming project and how to incorporate new methodologies into it.

Once a functional specification is available, detailed analysis of the proposed design begins. A test plan is generated for the device, it is usually broken down by major sub-blocks on the chip. Once the test plan is created, the actual tests are created. In some cases, special block-level only tests are created. However, in general, tests are usually created to run as much as possible at a full device level.

The following figure 1-1 represents the general verification flow in any design cycle parallel to the hardware development:

Figure 1-1. General Verification Flow

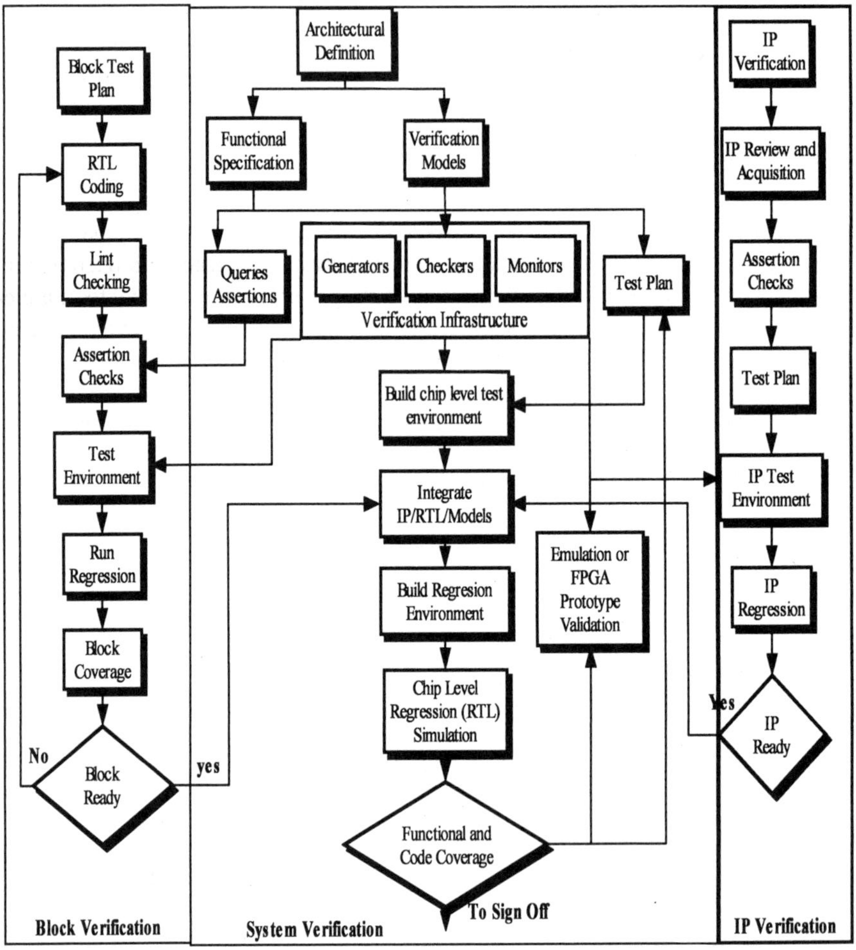

During the architectural phase the main verification tasks include:

- Develop verification methodology suitable for a system with most of the IP integrated blocks. Mainly the methodology is to write C, C++ or HVL (Hardware verification language) models of various blocks that can be used for plug and play while running system level simulation based on availability of the IP or block. Verification infrastructure should support reuse and integration of any such models built in house or acquired as verification IP.

- Figure out resource requirement in terms of CPUs, memory, man power, tools that will assist in verification.

- Based on tools/language used, detail the methodology to write verification models in a way that they can be reused as checkers during chip verification phase

- Start to develop system level verification infrastructure of stimulus generation and monitors at chip level interface and each of the IP/block interface that can be reused for individual IP verification

- Start writing down high level test plan for system level and IP verification

Once a functional specification is available, verification tasks at this stage generally include:

- Compose detailed test list from the functional specs for system level and IP verification

- Build test cases for the system verification using verification infrastructure.

- Build test cases for IP verification. Try to utilize existing test environment from IP vendors with a wrapper or a set of wrappers to integrate to our own verification infrastructure. Mainly requirements from IP vendors will be to provide functional/cycle accurate C/C++ model for the IP and also for stimulus generation.

- Build regression environment and run regression using IP/RTL/Models at system level.

1.1.1 Challenges

This section addresses the top challenges faced by a verification team. Some of these key challenges faced by design/verification engineers to get the SOC fully verified are listed below.

- When to stop verification? This is everyone's challenge and there is no well defined measure it. Companies define their own criteria like rate at which bugs on RTL is found, tapeout schedule and functional coverage.

- To have an integrated environment that both design and verification engineers can use for system level or block level testing. The challenge here is to keep up with the robust designs and changes/additions to features that keep happening within the same or different projects. The test environment should be reusable and independent to accommodate any such modifications and additions required.

- Since the simulation is limited by number of clock ticks, it is difficult to achieve an effective performance testing and statistic testing.

- Have a comprehensive tests plan and have an automated regression process that submits thousands of tests to all possible CPU resources concurrently, and automated reporting of each tests and bugs for failures. Reporting should also be able to write some nice graphs that show statistics of tests (passed, failed, crashed etc.) and for the bugs over time. From these statistics you can try to find out when to stop and sign off.

- How to define "Correctness" of the design?

- Developing tests that verify the correctness of the design, for example self checking randomization.

- How to measure effectiveness and improve the overall "Coverage"? Did my tests cover all of the interesting problems and if not, what needs to change?

- Need to debug at the next level of abstraction. Designs are getting too large and it is too time consuming to fully verify the design at the "signal" level.

- How to integrate different design teams' RTL blocks? As blocks get promoted into the overall design hierarchy, the tests must be expanded to ensure that these blocks are working correctly with each other.

- Some IP may be in Verilog, VHDL and/or C++. The Verification Engineer's challenge is to create a test environment for mixed language verification.

- How to validate the test generation effectiveness? As the model grows, are the functional tests robust enough to validate interesting corner cases?

- As the models change, how to maintain/track correctness? If the functional changes occur, how are the tests developed so that they can be re-usable?

There is still a lot to be done by EDA tool vendors to overcome these challenges and make verification simpler and automated process with some uniform methodology.

1.2   Topics Covered in the book

Various chapters in the book are arranged to provide the necessary concepts for effective functional verification of SOC.

1.2.1   Introduction

This chapter provides an overview of the book and the summary of all the chapters. It also discusses the reference designs that are used in the book. An overall verification flow and list of key challenges is also provided.

1.2.2   Verification Process

This chapters lists down various verification processes and tools available in the industry to aid functional verification. The idea is to be familiar with the possible solutions out there and choose the best that fits the design requirements and verification methodology. It talks about topics like - Typical verification Flow, Lint Checking, High level verification languages, Documentation, Scripting, Revision Control, Build Process, Simulation and waveform analysis, Bug tracking, Memory modelling, Regression Running, Code Coverage, Functional Coverage used in different verification flows and the tools that help functionally verifying the chip.

1.2.3   Using SCV for Verification

In recent years, engineers have adopted HVLs (High level verification languages) and the advantages are clear. In October of 2002 the OSCI standards body adopted SystemC Verification Extensions to improve verification productivity. This chapter will discuss what is needed for verification, why the facilities of SCV were picked, and how SCV can be used in your designs to improve productivity. The SystemC Verification extensions were created to increase the productivity of verification engineers. Verification engineers need many of the capabilities found in a pure software environment, such as pointers, abstract data structures, and object-oriented programming, in order to kept pace with design schedules.

1.2.4   Test Plan

The major reason for inadequate testing is an inadequate testplan which specifies requirements that are unclear, too general and have an unrealistic schedule. Every company has its own formal guidelines to create testplan from functional specification. This chapter gives an example of an effective test planning strategy. Test plan can be split into System level plan/methodology document and test description document. The methodology document can highlight the main features to be tested and methods to test them. Test description document lists the tests and maintains the coverage metrics until the sign off is met. The example has been provided for some of the cases from the network chip reference design.

1.2.5 Test Bench Concepts using SystemC

This chapter defines various testbench components necessary to build verification environment and give examples of how to build them efficiently using systemC. Whereas the methodology chapter focuses more on the overall verification environment for networking chip, this chapter goes in detail for creating testbench using SystemC language and examples from multimedia reference design. It is useful for the reader to get snapshot of both the environments with the examples. Testbench components are fairly well known these days. They typically include tests, test harnesses, monitors, master, slaves, etc. For transaction based verification, a method must be chosen to represent the system under test, injecting stimulus, and analyzing the results. All of these are dependent on each other, and they all require both tools and methodology to work efficiently.

1.2.6 Methodology

This chapter discuss functional verification methodology adaptable to any networking design with examples taken from the networking reference design discussed in the introduction chapter. The methodology gives examples of a self checking verification environment developed using Testbuilder (initial version of systemC verification extensions). Various methodology topics are extensively covered in this chapter like - Verification approach, Functional verification tools, Transaction based verification environment, System Level verification, Block Verification:, Block TVMs at system level, infrastructure, Stimulus/Traffic Generation, Rate Monitors, Checkers and Memory models. It also includes appendix that has code for some standard interfaces. The code is based on actual development work which was done in Testbuilder.

1.2.7 Regression

Large and complex designs are dependent on billions of simulation cycles to be fully verified. With the limited clock ticks simulators can handle, it becomes essential to set up the regression utilizing maximum resources and automation for concurrent test running. This chapter highlights all the essential features of automated regression setup and focuses on various trends for running regression, how to setup regression and maintain the regression based on your design requirements. Regression running should be an automated process where you start the regression from a start button and after that daily emails and reports get generated until all the regression is complete. Usually, this is not the case because there are many tools, jobs, memory space, machine issues which come in the way. So, it is required to do an estimation first on your resources for licenses and CPU slots and set up the regression based on design and functional coverage needs.

1.2.8 Functional Coverage

This chapter addresses the topics - Functional Coverage definition, Coverage tools used today, Functional coverage implementation and Functional coverage examples. Functional coverage is a new term that is appearing in the Electronics design industry. Traditionally we would all instinctively build functional coverage into our verification environments without thinking about it. Functional coverage is required since now we rely more and more on constrained random testing which allows tests to be directed at exercising functionality while randomizing some of the data in the tests. Since data is randomized there it is no longer sufficient to equate the execution of a test to a specific function that was exercised since information is generated on the fly. This creates the need to monitor activity during simulation and determine if the function was exercised.

1.2.9 Dynamic Memory Modeling

One of the performance challenges in functional simulation is simulating large memories in the design that take too much simulation memory space. This chapter focuses on providing a solution for using memory models for optimum simulation performance. These memory models can be either bought from memory solution providers or they can be built in house by designers. Topics covered in this chapter are - various alternatives for using memory models in simulation, running simulations with memory models, building dynamic memory models in house, buying commercially available solutions, advantages/disadvantages of different solutions available, dynamic memory modeling techniques and examples of dynamic memory models using test builder

1.2.10 Post Synthesis/Gate Level Simulation

Once functional requirements for design are met, timing verification is essential to determine if design meets timing requirements and better design performance can be achieved by using more optimized delays. This chapter focuses on dynamic timing simulation technique that can verify both the functionality and timing for the design. It involves exhaustive simulation on back annotated gates. Topics covered in this chapter include:Need for Gate Level Simulation, Different models for running gate netlist simulation, Gate Simulation with Unit Delay timing, Gate-level simulation with full timing, Different types of Simulation, Stages for simulation in design flow, Stimulus generation, Back annotating SDF files, Multicycle path, Using Sync Flops, ncpulse and transport delays, Pulse handling by simulator, Setting up the simulation, ATE vector Generation, Various ATE Stages, Generating vectors, Testing ATE, Problem at speed vectors and Setting up for ATE vector generation in simulation environment

1.3   Reference Design

This book uses two reference designs. One is the networking chip developed by "Azanda Network Devices". The examples are given in various chapters from different versions of Azanda chips that were taped out or were at the development stage. There might be some alterations from exact feature set supported to make the example look simpler. The other example is the multimedia Design Platform.

1.3.1 Networking Traffic Management and SAR chip

The reference design considered in this Section is Azanda's Schimitar device. It is full duplex, OC48, Traffic manager/ATM Sar device. The chip, uses OC-192 technology to provide high density low pressure solutions for OC-48 market. Supporting 1 million simultaneous flows and 256 Mbytes of external memory, the chip is truly large, complex and very difficult to validate in a simulation verification environment. These are the main features of the design:

- Open architecture - works with existing NPU architectures and code bases

- Dual-function Traffic Manager and ATM SAR works on the line side as well as the fabric side of the line card

- Applies Azanda's 10 Gbps, 0.15 micron technology to high-density OC-48 (and below) solutions as well as OC-192 solution

- Currently sampling with volume production in early Q4'02

ScimitarTM Device as SAR

- Terminates ATM

- ATM to Packet Networking

- ATM Classes of Service and QoS

- Policing, WRED, Queuing, Shaping and Scheduling

Scimitar as Traffic Manager
- Manages queuing and scheduling
- Switch fabric segmentation and reassembly
- Shaping and scheduling on egress

1.3.1.1 Design Overview

To understand the functionality of the chip, lets walk through the block diagram shown in figure 1-4. Packets or cells are received from either the line or the switch fabric at the incoming interfaces. The data proceeds to lookup engine, where one of the several lookups is performed, including an ATM lookup, a multi protocol label switching (MPLS) packet lookup and an MPLS-like tag from an network processing unit or classifier to identify the flow.

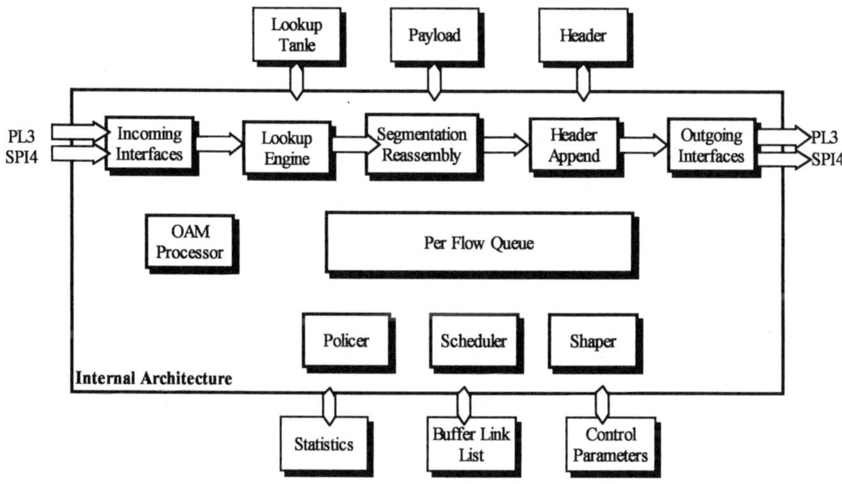

Figure 1-2. Block diagram

Next, the segmentation and reassembly engine, with the help of external data memory, segments packets into cells, reassembles cells into packets or propagates packets or cells. External data memory is used to enqueue the packet or cell payload until it is scheduled for dequeue onto an outgoing interface. The header append block adds the outgoing header onto packets or cells, which could be either ATM cells or cells specifically configured for the switch fabric. The header append block also does MPLS processing such as tag translation and push/pop. Finally, the outgoing interfaces send the packets or cells out of the device to the line or switch fabric.

While the data path performs those tasks, the per-flow queue keeps track of all the data stored in data memory on a per-flow basis. The traffic shaper and output scheduler manage the activity of the per-flow queue and determine how to schedule the packets or cells out to the outgoing interfaces.

1.3.1.2 Schimitar Technology

Hardware implementation

- Deterministic line rate performance from 2.5G-40G
- highly software configurable

Multi Service Architecture

- supports cells and packets
- supports inter working between ATM and packet networks

QOS support

- support CBR, VBR-rt, VBR-nrt, UBR+, GFR
- Supports packet based SLAs

Flexible scheduling and shaping algorithm engine

- multiple levels of cells and packet aware scheduling
- supports flow aggregation, tunnels and VOQs

Best of breed approach

- Industry-standard interfaces for easy integration
- Works with any NPU or switch fabric

1.3.1.3 Schimitar Capabilities

Dual traffic management roles

Fabric Traffic Management Specific

- VOQ scheduling and flow control
- WRED and class based bandwidth management
- Header add/delete features
- SAR for the switch fabric Standard interface to the fabric with flow control SPI4.2 or PL3

Line Traffic Management

- Manages up to 64 line side ports
- Port based and class based bandwidth management
- ATM or MPLS classification and header modification
- Fine-grained, packet or cell shaping and policing
- ATM SAR functionality
- Standard interfaces to the framers

1.3.1.4 OC48 TM/SAR Feature Highlights

- 256 MB of ECC protected payload memory
- 1M simultaneous reassembly and segmentation flows
  - Per flow queuing, buffer management inc. WRED
  - Per flow statistics, RR/DRR/priority scheduling
- Traffic Shaping/Policing
  - 32K on-chip, extensible, traffic shaping/policing resources
  - 4K traffic shaping profiles
  - 1K traffic policing profiles
  - 2 Timing wheels with 64K slots each
  - Aggregated shaping (Shaper Tunnel).
- Flow aggregation: 16K roots, 128K leaves
- Virtual Output Port Scheduling with back-pressure
- Multicast: 1K roots, 128 leaves per root

1.3.1.5 Device Characteristics:

- Wire-speed Traffic Manager
- Integrated SAR
- Full-duplex OC-48 line rate
- Three full duplex datapath interfaces
    - POS-PHY Level 3 PHY
    - POS-PHY Level 3 Link
    - SPI 4.2
- Chip statistics
    - 40mm x 40mm BGA,
    - <5 watts, 1.2V core
- Subsystem requirements
    - 3000 square mm
    - <15 watts for 128K flows

1.3.2 Multimedia Reference Design

The multimedia reference design is a digital picture frame. The design takes an encrypted JPEG image in from its Ethernet port, decrypts the image, performs a JPEG decompression, and sends the output to the VGA display.

1.3.2.1 Design Overview

The design contains two processors, one for control functions and one to perform the DSP function of JPEG decompression. It also contains two Ethernet ports, a DMA block, a Decryption block, and a VGA block.

The Multi-media reference design provides the following:
- Capability to send and receive data via
    - 10/100 Ethernet Port,
- Capability to drive a VGA monitor
- 56 bit data encryption capability
- A programmable platform that provides
    - Controller capabilities
    - DSP capabilities

- Firmware for JPEG decompression

The Ethernet port supports both 10 mbit and 100 mbit operations. It feeds into a four channel DMA block using FIFO's for flow control. All of the blocks are connected to an AHB bus. The software executing on the micro-controller takes the Ethernet packets and uses DMA transfers to send the data to the rest of the system for processing.

The design within a system configuration is shown in Figure 1-5.

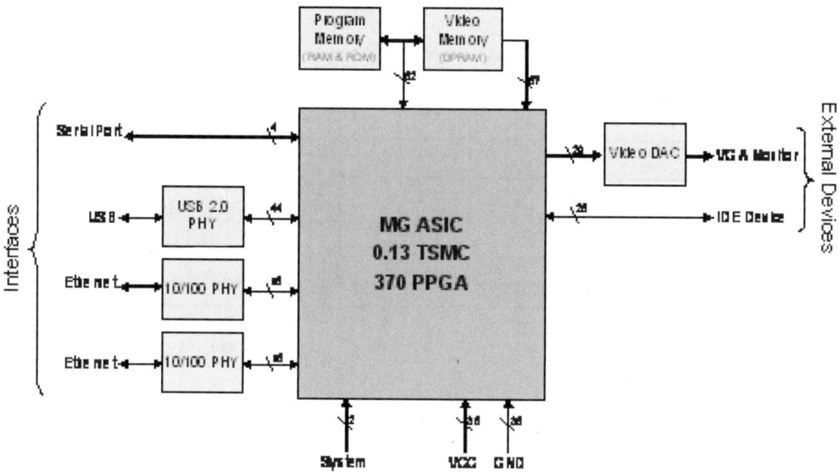

Figure 1-3. System configuration for multimedia design

The functional blocks within the design are shown in Figure 1-6. These components provide the functionality to receive encrypted JPEG data through industry defined interfaces (Ethernet), decrypt the JPEG Data through the DES Block, decompress the JPEG file and display through the VGA interface.

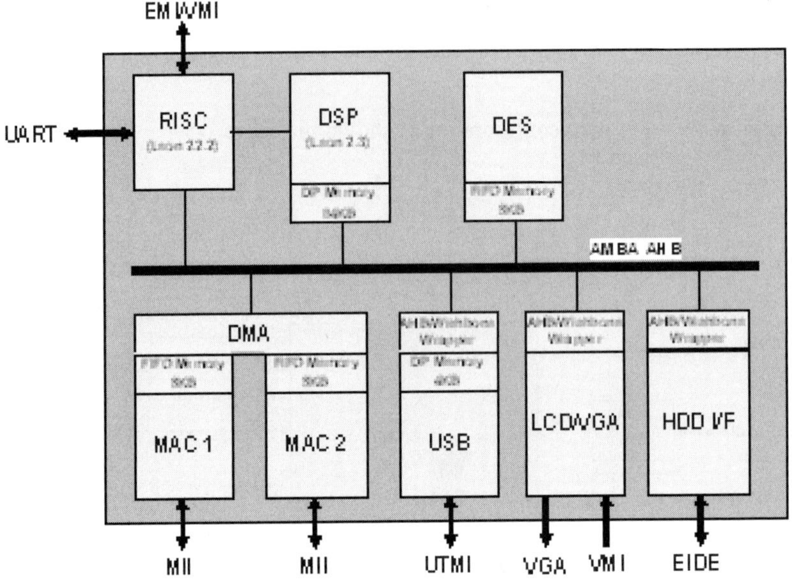

Figure 1-4. Block diagram

CHAPTER 2

Verification Process

To start any system verification there are three main questions to be answered first:

What to verify?

What tools, processes to use?

How to verify?

System specifications derives the verification strategy that is followed for particular chip verification. These specifications along with the test plan answers what to verify. Verification processes answers what tools and processes to use, and verification methodology answers how to verify the chip.

Verification methodology and implementation is dependent on the set of available tools and the process adopted for verifying the chip. This chapter discusses different tools and processes that are available to be used in functional verification of a chip. It is essential to be familiar with various verification processes to be able to choose the most appropriate ones for the methodology that will be followed for verifying the chip.

Topics covered in this chapter:

- Typical verification Flow.

- Lint Checking.
- High level verification languages.
- Documentation.
- Scripting.
- Revision Control.
- Build Process.
- Simulation and waveform analysis.
- Bug tracking.
- Memory modelling.
- Regression Running.
- Code Coverage.
- Functional Coverage.

2.1   Introduction

Extensive, thorough and rigorous functional verification testing is the requirement for complex architectures of today's designs. Design scope is too complicated to rely on old methods. So, the best verification strategy is the one that starts early in the design cycle concurrently with the creation of specifications of the system. Verification now requires transaction verification methods and technologies to capture tests and debug the designs.

Transaction level verification provides the solution to meet upcoming challenges in functional verification because of increase in gate count and complexity of System on chip designs, ASICs and FPGAs. Test stimulus generators, transaction protocols, monitors, protocol verifiers, and coverage monitors are captured at the transaction level.

Typical Verification Flow for any design is as in figure 2-1. Verification team starts working at the beginning of design phase when function specs are being developed. The test plan defines the design configuration based on the functional specs. It also defines the stimulus to be applied to the design, expected response, functional coverage criteria, and the tape out requirements. The next stage after the test plan is developing the verification infrastructure of stimulus generation, checking, monitoring and regression. Once infrastructure is developed, a test bench is generated based on detailed functionality and categories to be tested in the test plan. There are deterministic tests for specific corner cases and typical scenarios and there are constrained random tests to cover random scenarios. Once all the tests are ready, they are run in regression with all different configurations defined in the test plan. Test response is checked based on automatic checking or post analysis of report files generated from the regression run. Automatic response checking has monitors at all the interfaces that pass on the information to dynamic checkers running in various threads on simulation. If response is incorrect, it is further analyzed and debugged and a bug is filed in case of RTL problem. Once all the tests in test plan are run, next step is to meet the exit criteria based on functional coverage or bug rate metrics or any other metrics defined for tapeout. After months of running regression and very low bug rate, if exit criteria is met for the tapeout, the verification can sign off.

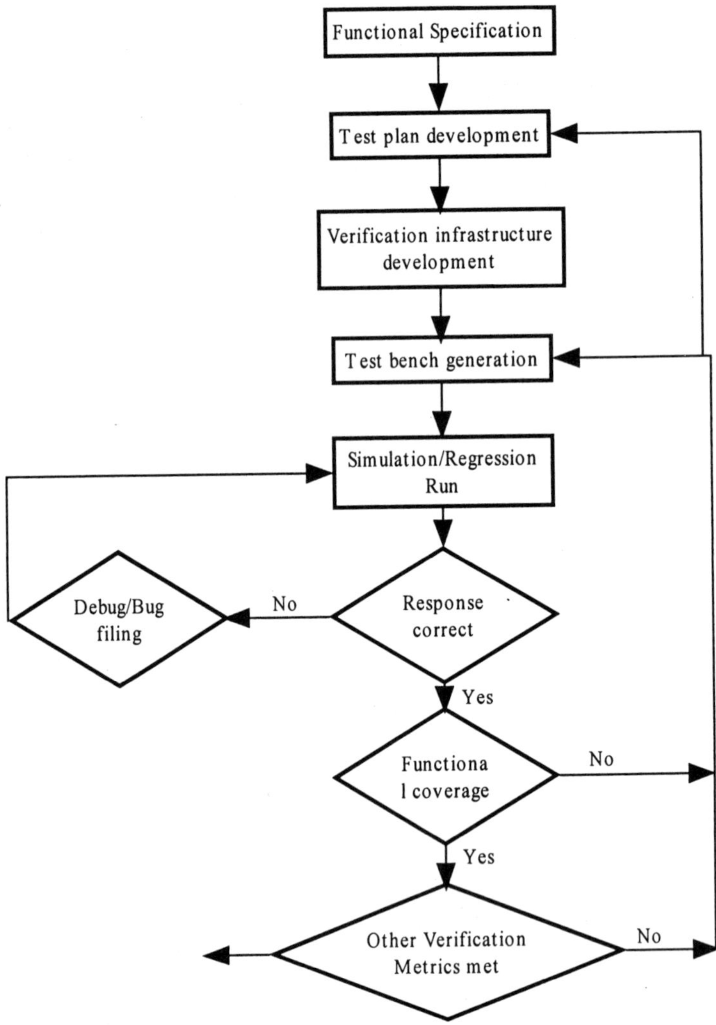

Figure 2-1. Typical Verification Flow

The following sections describe various verification processes in detail along with the verification tools that are mostly available and used.

2.2 Lint

Lint tools are widely used to check HDL syntax before going to synthesis and RTL simulation. They purify Behavioral and RTL code to assist RTL designers. Designers can choose graphical user interface or batch mode to filter which errors they wish to view by type, category and severity. These tools have built in lint checks to locate and syntax or semantics errors in source code. They also provide a front end in some common scripting language, for example perl. This allows designers to build their own lint checks suited for their own unique design environment. Lint tools are very popular in designing silicon chips, they are simple but essential for RTL purification.

Features:

- Graphical mode as well as batch mode supported.
- Learning curve should be low.
- Tool should be capable of handling thousands of lines of code without impacting the performance.
- It should support all the syntax in Verilog or VHDL language. It should not require another modeling style to be able to use the tool.
- processed code should be able to simulate and synthesize.
- Should be compatible with popular simulators and synthesis tools used in the industry.
- Gives user enough control on filtering the error messages instead of dumping too much information.
- Can handle FSM extraction and race condition detection.

Many vendors provide lint tools, their checking features overlap in some areas and differ significantly in others.

- SureLint: It is quite common static design analysis tool for analyzing and debugging complex designs before simulation and synthesis. This offers FSM analysis, race detection, and many additional checks. It allows user defined checks to add to library of built in checks.
- HDLLint: Configured Lint tool for Verilog and VHDL. Comes with Perl front end to allow designers to build their own rules.
- ProVerilog: Linting tool from Synopsys for Verilog comes with predefined rule sets to check. Designers can program their own set of design rules to check using Rule Specification Language (RSL).

- ProVHDL: This Linting is also from synopsys for VHDL.

- Nova-Verilint: Configurable Lint tool for designers using verilog. It uses the same command line as Verilog-XL.

- SpyGlass: Support for both Verilog and VHDL, comes with rich suite of built-in rules. Custom rule development interface provides C-level access to RTL. Also, has perl interface to help integration into design flows and custom report generation.

- Incisive HAL: HDL Analysis tool from Cadence designs systems.

- VN-Check: Configurable Lint tool for Verilog and VHDL fromTransEDA. Comes with comprehensive set of pre defined rules. Has powerful GUI interface.

2.3   High Level verification languages

Most of the time adhoc verification methods are used because of the time pressure to tape out the chip. Using these adhoc methods is a major obstacle which comes in the way of being able to generate reusable verification environment. It is required that functional verification flow be standardized to get rid of this major hindrance to reusability of verification environments. The most important element in any standardized automatic functional verification flow is the language. If all verification environments use a common language, then different verification groups can talk and share their expertise. This will help to bring out the best solutions to common verification problems and reduce the effort in trying to reinvent the wheels by using different concepts from different languages.

For complex chips the transaction level verification method is most suitable. It is possible to implement a testbench for transaction level verification using available high level verification languages. The testbench created using high level verify languages is compact and easy to understand.These languages also offers enhanced productivity to verification engineers by reducing the problem and the time spent in developing complex test benches. They also have support for interfacing with HW/ SW coverification tools. These languages have features to develop verification components that are otherwise impossible to build using traditional verification components. These components developed in high level language can model some standard interfaces or protocols stimulus generation, monitoring and checking. They can be used from design to design without any modification at all.

There are several verification languages used in the industry which have technical abilities for advanced verification methodologies. These are available in following two categories.

- Open source, mainstream language solutions: SCV (SystemC Verification extensions), Testbuilder (initial version of SCV), TestWizard(AVery).

- Proprietary verification languages: SystemC, Specman(Veirsity), SystemVerilog, Vera(Synopsys), Superlog(Co-Design), Rave(Forte).

Testbench written using c/c++ rely on being able to use high level data model to represent transaction type classes and user defined data structures. Records and structures in standard c/c++ are much more robust. Records can be hierarchical and express linked lists. They can support powerful search facilities by record type, field value, index and time. Most of the Proprietary verification languages also provide these useful features. For example, Specman data structures are hierarchical by default, and can do powerful searches on all the search types listed.

A good high level verification language must have following features.

- High level data types. Each data or control word element should be able to represent in the record structure unique element.

- Lists and collections that can allow to enumerate field value sets and aggregations.

- Random generation features to be able to write constrained random cases.

- Support for temporal expressions. Rules for protocol dependences and checking are expressed using temporal expressions.

- Functional coverage support for evaluating range of parameters and features exercised during simulation.

Assessment of any language comes down to evaluating the following:

- Native verilog extension vs. new language: Proprietary languages are usually extension of existing veriliog features. Many basic constructs and types overlap in these languages. They are known for early adoption by the design teams, who are unfamiliar with C/C++, and cover most of the basic needs. For supporters of C++, they find it much easier to architect advanced verification testbench by making use of function overloading, stl maps, templates etc.

- OOP vs, Procedural programming

  Object oriented programming model definitely makes advanced transaction based verification easier to implement. Most of HVLs also provide some OO features. Verilog HDL modules and instances are always static unlike general OO languages which support dynamic object creation and destruction. General OO languages also have derived classes, virtual functions and classes, class and functions templates, and exception handling etc. Users can develop their own classes and methods, and use all power of C++ programs for testbenches.

One can model non-determinism more flexibly in HVLs than in an HDL by separately scheduling execution threads and randomizing input values. In addition, one can vary a directed simulation test and track the different variations that are created, so that it can be determined how much coverage was obtained from any given simulation run. Use of C++ based language or any other proprietary verification language enhances HDL capabilities with object-oriented coding, data structures, reentrancy, and randomization, while preserving HDL mechanisms such as parallel and sequential execution and event and delay control.

- Testbench automation features: All these languages should have good support for automation features, search capabilities, functional coverage analysis and query functions.

- Most block level verification environments do not require elaborate transaction verification methods. Language should give flexibility of choosing a simpler solution or methodology.

- For the designers who have mostly dealt with Verilog or VHDL, learning curve is involved in getting used to object oriented concepts of the HVLs.

- Reusability: These languages should be able to provide reusability of various verification components. Like interface protocol written once for monitoring, generation or checking should be able to hook up anywhere. Also, stimulus generation task should be able to generate stimulus on any interface of the chip without changes in the code.

- Some languages have ability to communicate between various simulations using socket interface. This is essential for large distributed simulations. One chip can be simulated on system a and other on system b, and information can be exchanged back and forth between the two at much higher speed than simulating both chips on one system.

- Semaphore, mutex, multiple threads running at same time.

- Self checking testbenches should be easier to do.

- Is tool open standard, with support from EDA vendors?

- Is this a point tool or part of integrated verification solution? Can it cause simulator to slow down because of inefficient PLI calls?

- What does it take to be productive with the tool?

- Does verification language resemble with well known languages - c, c++, Verilog or VHDL?
- What is the performance of the tool?
- How powerful is debugger at handling multiple threads?
- Is the code structured and easily understandable?
- Is there additional software to worry about, additional training required etc.?
- Support components integration from different languages, in case IP is in a different language.

Most of the verification languages try to provide the required capabilities listed above. Some are better than other in some cases. To explore more on what kind of features are provided with these HVLs, lets take an example of SCV.

As soon as verification complexity increases, the comprehensive C++ testbench class library that extends the capability of an HDL simulator starts coming in handy. It provides object-oriented techniques to help author complex testbenches efficiently. Using the classes and methods in the SCV library, one can create sophisticated concurrent transaction-level tests for a design. Initial time to be spent in hooking up SCV is generally one day's work. The testbench development problem is transferred to the C++ domain. Data structures provided in the SCV library that is STL based can be used. C++ talent available can be used to define the chip's own data structures and use all extensive C++ language features. Running multiple threads simultaneously and being able to use semaphores and mutex is a big help. The spawn feature is extremely useful. Random complex situations can be described in a concise manner.

To supplement the existing capabilities of HDLs, SCV provides the following features for developing more complex tests:

- Data structures.
- Dynamic memory and tasks.
- Re-entrancy and recursion.
- Support for object-oriented programming.

Because the SCV class library is written in C++, it provides C++ features such as strong type checking, code reuse, pointers to functions, dynamic memory allocation, dynamic spawning of new threads of execution, and object-oriented features such as overloading and inheritance. SCV allow the use of available C++ tools such as profilers and debuggers, to improve C++ testbenches

Transaction verification models (TVMs) can be created, that spawn concurrent tasks. This capability makes it easier to develop self-checking cause-and-effect tests. This type of testing uses active-slave TVMs connected to the output ports of the device under test. SCV supports constrained random test generation through a variety of mechanisms:

- random number generation management, task profiles that define the constraints for randomly generating task arguments, and TVMs.
- Other mechanisms include task groups that allow for the definition of "bags" from which specific TVM instances or TVM tasks can be chosen, according to a dynamically-controlled weighted random selection process.

It automatically records every task invoked from test to TVM, including

- Task type, such as do_read.
- Task arguments, such as address and data.
- Relationships between cause-and-effect transactions on different interfaces.

Simulation debugging is simplified dramatically by automatically recorded transaction-level debugging information. This information can be viewed graphically in a transaction-viewer.

Also, it automatically records the information needed to perform functional coverage analysis in the simulation database.

2.4   Documentation

There are tools available for documenting, understanding and maintaining impossibly large and complex code. Before evaluating any of these tools, it is necessary to consider the documentation needs and choose the tool that best suits the requirements. Besides documenting the infrastructure and testbench code, appropriate document tools should be selected for writing testplans, flow charts, diagrams, reports and presenting metrics. Some examples are Framemaker, MS Word, Wordperfect, Adobe Acrobat, SmartDraw, CorelDraw etc.

Features for documentation tools:

- It should be simple to use. With some tools, documentation can consume more effort than project development.
- It should be integrated with testbench and infrastructure development, it shouldn't be entire separate project.

- Multiple people should be able to contribute to the documentation effort.

The following are some of various commercial and non-commercial documentation tools available:

- Synopsis: Cross language documentation generator written in Phython, supports C++.

- AutoDoc: Documentation generation for Tcl.

- ROBODoc: Multi-lingual documentation tool supports C, C++, Java, HTML, Tcl, Shell scripts and others.

- AutoDuck: It extracts embedded documentation from C, C++. Output formats include HTML and RTF.

- Doxygen: Embedded documentation for C, C++. Special commands can be embedded in marked comments.

- DOC++: Documentation tool for C, C++, Java - can output HTML and LaTeX.

- Framemaker: designed to produce printed media, but not very good at online help.

- Robolhelp: Used for producing HTML documents but user needs to be an expert in HTML.

Some commercial documentation tools are DocBuilder, DocJet, Doc-o-matic, Genitor Surveyor, CC-Rider etc

Take an example of Doxygen tool for extracting documentation from C++. There are very few special commands to remember while commenting the code and the process of generating documentation can be automated along with testbench build process. It is simple to learn and produces very well formatted documentation. Following section shows an example of how to set up documentation for the code.

2.4.1 Documenting verification infrastructure

This section describes the necessary details to document the verification infrastructure code using Doxygen. Out of the various features provided by Doxygen to document the code, the following are the main four points to remember:

1. Use "///" for commenting classes and members before they are actually declared.
2. Use "\a" to point to actual function or task defined.

3. Use "\ingroup" and "\defgroup" to group together (not necessary the first time of documentation release).
4. Use "\param" to define function parameters.

Steps

In <*.h> files (for example any /projects/verif/doxy/tbvChecker.h):

- Describe every class

In the above file class tbvCheckerTvmT can be described as:

1. /// Top level class for implementing checkers.
2.
3.
4. /// This class provides implementation structure for creating checkers.
5. /// It provides simple queuing service and functions to inherit from.
6. /// Inherited checkers should overload the \a process(), \a compare() and \a locate()
7. /// functions.
8.
9. /// \ingroup checker

Notice that each of these lines has three backslashes. Use three /// for doxygen to recognize the comments.

Line 1 is taken as brief description of the class.

Lines 2,3, are left blank to distinguish brief description from detailed description.

Lines 4,5,6,7 are detailed description of the class.

Line 9 is in case it is needed to group this in the group called "checker". It is a good idea to group different tasks and classes for clarification.

The group is defined at the beginning of <.h> file as follows:

1. /// \defgroup checker checker parts.
2. /// Associative parts of checker implementation.

Line 1 defines group checker and gives brief heading "checker parts".

Line 2 gives more description to it.

After defining a group, one can add to this group by \ingroup checker. In tbvChecker.h, tbvReceiveTask(task that receives data) and tbvCheckTask(task that implements check algorithm) can be added to this group.

Notice the "\a" in front of function names in line 6. In the documentation this will take you to the actual function call description, in case "\a" is attached in front of the tasks or function names.

- Describe every member in <.h> file

For example in the above file:

/// used to serialize the DataStream from the receive channel.
tbvMutexT   receiveMutex;

/// used to serialize the datastream from the send channel.
tbvMutexT   sendMutex;

These members have brief description now.

- Describe functions and parameters in <.cc> files

Example tbvChecker.cc
Function process in this file is described as follows:

1. /// called when a new cell comes into the input queue.
2. /// This function is called when \a tbvReceiveTaskT receives cell
3. /// from monitor.
4. /// \param key Index into queue that new cell is located in waiting to be processed.

void tbvCheckerTvmT::process(const tbvSmartUnsignedT & key)

Line 1 is brief description.

Lines 2,3 are detailed description pointing to the actual tbvReceiveTaskT.

Line 4 uses "\param" to describe parameter "key" of the process function. This way all the parameters can be defined.

Remembering this information while writing the code, can automatically generate quite good documents using Doxygen. To get more information on Doxygen and use more features, go to following web page:

http://www.stack.nl/~dimitri/doxygen/manual.html

As an example, one of the outputs from Doxygen is for dataCom class defined in testBuilder(initial version of SCV)as shown in figure 2-2, Inheritance diagram for class tbvDataComT:

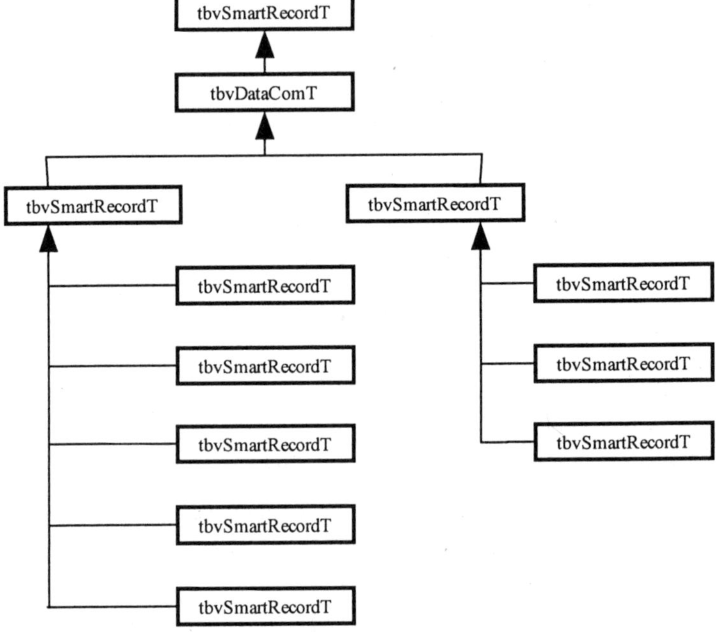

Figure 2-2. Inheritance diagram

2.5   Scripting

Scripting expertise is required for verification environment to deal with regression setups, reporting, metr.ics and result analysis work. Scripting languages are also called gluing language or system integration languages. They are generally not used to build whole test bench from scratch, their main purpose is plugging together various components of testbench infrastructure for regression and reporting.

Features to look for help in scripting tasks:

- Should not be system dependent, it should be portable over different Operating Systems.
- Should have simple debugging options.
- GUI functionality should be supported.
- Internet access support.
- Easier to write and understand.
- Should be stable language having no risk of backward incompatibility.

Difference in scripting and system languages. Both system and scripting languages are required for different kind of tasks.

- Scripting languages are higher level than system languages. A typical scripting language statement executes machine instructions in hundreds or thousands as compared to less than ten for system programming language.
- Scripting languages are much easier to learn because they have simpler syntax and there are no complex features like objects and threads.
- Scripting language uses interpreter so it is less efficient than compiled code of system language. There are some scripting languages like Perl that can be compiled and save performance.
- Primitive operations in scripting language have greater functionality.
- Scripting languages are typeless whereas system programming languages are strongly typed.
- Strongly typed nature of system programming language discourage reuse.
- Scripting language do the error checking at the last possible moment when value is being used. System languages can detect errors ar compile time.
- Generally system programming language is more code and less flexible programs compared to scripting language but they are highly efficient and can run at 10-20X faster since there are fewer run time checks.

- Applications can be developed much faster with scripting language but for complex algorithms and data structures, the strong typing of system programming language makes it more manageable.

Here are some of the scripting languages:

- Tcl/Tk: This has almost are required features and easy to learn. Compared to Python it is not very applicable for large programs and is weak in data structure. This was designed to be used in applications. It has simple and clean interface with C.

- Python: It is open source and has almost all the required features. It is available for all important Operating systems, and is widely used. Negative is the indention and speeding.

- Ruby: This has O-O support better than phython and is very flexible. It can add method to a class at runtime. Not very widely used yet.

- Perl: This is widely used and is very easy to learn for engineers who are already familiar with Unix shell programing env. It has flexible set of control statements, powerful open statement etc.However, Perl files can be harder to read.

- Javascript: This has ability to be used as macro language and as regular shell and it has very clean syntax. This is embedded and can be used instead of TCL. It has more use in HTML.

- Scheme. Syntax is not very clean. It has several existing versions that are not compatible. It comes from LISP.

- Shell script. This is commonly used for Unix and is simpler. It is not portable, since it depends on external programs. ksh93 is actually first scripting language that is also a Unix shell.

- PHP: It is very useful for solving wide range of tasks in WEB enabled environment. It has seamless integration with MySQL database and Apache WEV-server.

2.6   Revision control

Revision control and bug tracking are two major aspects of verification processes. Version control is more used for development and bug tracking for collecting metrics. A good version control system should be easy to use especially in branching, should support multiple users who are working in chaotic cooperation, support existence of many different versions while working towards one central version.

Adapting to version control should be mandatory in developing any verification infrastructure for large design mainly to aid all developers for making changes in common database. There should be a process to manage multiple versions of verification database so that every verification engineer can

- Check out and work the working version of the database, and compile it without wasting time in debugging the overall code that is broken because someone else checked in his temporary changes.
- Can make his own changes to test out the verification infrastructure he is developing without affecting the work of rest of team members.

If everyone is just using version control tool and there is no common process to be followed to maintain it, there is the risk of infrastructure database going totally out of control because everyone is checking in their temporary changes.

One example - say if you are using CVS version control, and do not want to deal with complex branching, you can have an integrator responsible for tagging the whole environment once or twice every week and releases it to everyone. This way everyone can have access to latest checked in, but also the working version is always tagged and available to be used. It is like taking a snapshot and saving it to be used later on. Also, since this snapshot is from the checked in database, there can be more changes done to it and merged with the latest code. Someone's temporary changes will not affect others or break the design at any time.

Other solution is to use branching but that makes integrator's job little more difficult to merge everything together.

Below are some of the available version control tools to select from

Various version control tools

- Clearcase: Clearcase needs high maintenance and dedicated resource to take care of it. It is expensive too.
- Perforce: Is also moderately expensive.It is reliably faster and supports almost all the operating systems. It uses central database and have support for many platforms.
- VSS: is from microsoft and works in windows only. It is not known to be very reliable and data can get corrupted. It comes with repair tool, if something is buggy, repair tool can be used. There is no insight into the code - it is like black box.

- CVS is most common and widely used open source software. It is free and reliable. It is is file based, and not known to be stable on windows. Its client server access method lets developers access the latest code from anywhere there is internet connection. Unreserved check out model to version control avoids artificial conflicts common with exclusive check out models and its client tools are available on most platforms. Negative is - requires central server, branching is not very easy, rearranging of directory structure is not recommended.There is community site available for any CVS support http://www.cvshome.org

- Arch: It is newer but seems better than CVS and Subversion. It can allow each different user to create a branch and make it available to others. Negative aspect is that it is based on shell scripts and not portable.

- Subversion: It is open source software claiming improvement and almost replacement of CVS. It supports renames, symbolic links, atomic commits, easy branching and tagging etc. It is still under development.

- Aegis: This is another open source software to replace and improve CVS. It is more comprehensive and only for unix.

- Vesta: This is configuration management from Compaq for Unix in C++ which exists for about10 years. It is required to be used by professionals and has long list of commands. It uses a central repository that is mapped to file system.

- Bitkeeper: It is commercial tool that is recommended by many users.It can handles renaming of files.

- PVCS: One more of commercial configuration management tools used.

- SCCS: Source Code Control System. This is standard on Unix systems. Used only for local systems. Does not provide that much functionality.

- RCS: CVS is built on RCS, it is one of the old time revision control system. It is used only for local systems.

- Continuus: This is expensive revision control system and is quite complicated.

- PRCS: This claims to be easier to use, it almost works like CVS, SCCS, RCS

- Sourcesafe: This is commercial and only for MS-Windows.

- CBE:stands for Code Building environment that is written in Java which can be file based or user data based. This is also known to be used as a make system. Not many people are using this.

- Projectl: This is web based project management in Java. This includes many other functionality like task management, appointment, bug Tracking etc. It is still in development phase.

2.7   Build Process

Build is to create a program from sources it has to be compiled. There are many tools that exist to do this automatically on many different systems. Among various names of existing build tools are freeBSD, Portage, Autoconf, Automake, SCons(make tool in python), Cons, tmake, qmake, jam atc. For any verification related C++ testbench, it is recommended for build proccess to use GNU make,which can compute dependecies amongst the design to compile only that which is necessary to be recompiled. Compilations are also dependent on different simulations that will compute dependencies themselves. Testcases, external verification files and scripts will also be necessary to be compiled and run. All this can be done using make and other scripts. It is recommended to have option for both dynamic and static builds. With some of the existing verification tools, build process is built in.

2.8   Simulation and waveform analysis

Number one requirement in simulator is performance. Everything depends on simulator engine and simulator must execute efficiently at RTL level of design abstraction. Also, simulators have to meet the demands of increasing gate counts and higher levels of abstraction. Here are some of the common commercial simulators used.There are some free available but they are limited on number of verilog lines they support:

- ncsim, ncverilog and ncvhdl from Cadence is high performance simulator with transaction/signal viewing and integrated coverage analysis. This ensures accuracy with certified libraries from over 30 ASIC vendors. This is most commonly used. It handles gate level simulations very well.

- Polaris: It is verilog event driven simulator from Avanti.technologies.

- Speedsim: is verilog cycle based simulator from cadence for verifying digital logic designs.

- Verilog-XL: Old Verilog simulator from Cadence.

- VeriLogger Pro - It combines features of traditional simulators with graphical test vector generation. It is good for unit level testing.

- VCS: Verilog simulator from synopsys. This was first alternative simulator for Verilog and is used widely.

- Scirocco: VHDL simulator by Synopsys specifically designed to meet unique requirements of SOC verification. It integrates cycle and event based simulation techniques in single simulator.

- VHDL Similli from Sympony EDA is event based VHDL simulator assembled in Integrated development environment and waveform interface.

- ModelsSim: It is mixed language event driven simulator. It is lesser expensive, has good debugging capabilities and User interface. This is a compiled simulator and has comparatively good performance.

- FinSim: From Fintronic also supports many different popular platforms. It is verilog event driven and cycle based simulator. This is compiled simulator like ncverilog and VCS but slower.

2.8.1 Waveform Analysis tool features

- Can display live data as it begins recording without much performance hit on simulation run as well as post simulation data loaded from the database.

- Allow to combine signals to form buses, conditions and virtual signals.

- Allows to search objects in design without regard to design hierarchy.

- Tcl commands allowed.

- Markers, flags allowed. Should be able to measure simulation time.

- Should be able to define groups.

- Should be able to define expressions that perform arithmetic or logical operations.

- Good to have schematic tracer option to show HDL design as schematic diagram and help to trace signal through the design.

- Should be able to watch values of signals as they are changing in simulation.

- Should be able to trace back the selected signal to source code where it is defined.

- Allows to store database in compressed form.

- Allows limit on the waveform size.

Following are some of the waveform tools available that have above features:

- Debussy: FromNovas - Waveform viewing and analysis tool for fast tracing of design structure.

- Signalscan: originally from Design Acceleration Inc. It is widely used waveform viewer with highly interactive environment for waveform analysis.

- WaveformerPro: From SynaptiCAD waveform viewer, comparator and analysis tool.

- Simvision: From Cadence, graphic environment to simulate, debug an analyze the designs.

2.9   Bug tracking

Bug tracking is one of the metrics used for verification signoff. To be able to keep track of various bugs raised, resolved - an automated bug tracking tool is required that makes it simpler to analyze progress using metrics and reports. Select best bug tracking tool that fulfills your requirements. Freeware bug tracking tools such as Bugzilla bug tracking database, GNATs bug database, or Debian bug tracking system usually takes a long time to set up, maybe not user friendly or supported. Expensive bug tracking tools may have some special rarely used features that are not worth the cost. Cost of bug tracking software varies from free to thousands of dollars. Some free systems charge a heavy consulting fee for support.

These days bug tracking systems are available that are web-based and require no client software installation. In old bug tracking systems which are client based, their is requirement for installing client server and each user need to install client software. If various users are located remotely and connected through internet web-based bug tracking is best solution.

Following if the list of some of the features to look for in the bug tracking tools:

- Web based bug tracking system should support browsers your users are using. Some free available bug tracking systems do not work well with windows. Mostly free bug systems are linux based.

- Make sure that backend database is as required. Some free systems may only support just one database and not multiple databases. If it uses nonstandard database you should know of it beforehand.

- Programming language in which bug tracker system is written may be relevant in selecting bug tracker system based on your skill set.

- Bug tracking system should be lightweight, robust and fast.

- There should be minimal system requirements and maintenance.

- It should be easy to install and efficient to use.

- There should be no learning curve involved in recording or querying a bug.

- Should be platform and database system independent.

- Should support multiple projects and access control.
- Should allow for automatic bug filing or and manual bug entry.
- Should support file attachment.
- It should give email notification and support copying list for email.
- It should be able to accept bug report submitted through email.
- Reporting for metrics should be supported in different forms.
- Search and Query should be simple and extensive with support for searching combinations of all different fields.

Bug tracking tools:

Following is list of some of the many available bug tracking tools. List includes available free bug tracking tools and as well as some of commercial bug tracking applications to cover kind of feature list provided by these tools.

Free bug tracking tools

- Bug Tracking: Bug-Track.com: It is free bug tracking system that is 100% web-based, fast robust and simple to use. It supports email notifications, file attachment, history tracking, and advanced customizing.
- Bugzilla: bug tracking software form Mozilla project. It is written in Tcl and uses mySQL for database. Not very simple to setup and maintain.
- GNU GNATS/PRMS: It supports command line use, email, web interface etc. It is commonly used but requires expertise and dedicated resource to setup and maintain.
- Jitterbug: It is a web based bug tracking tool originally from Samba. It uses file system instead of database. It receives bug reports via email or a web form. It is implemented in C and requires web server.
- FreeTaskManager: Free web-based bug tracking tool.
- GranPM: It supports email notification, file uploading, employee time sheet tracking and has complex security.
- Teamatic: Free enterprise web based bug tracking system.
- AdminLog: Windows based running SQL server database. Does not have very robust security permissions.
- BugRate: Java software that supports bug reporting by web and email.

Other bug tracking tools:

- IssueZilla: It is web based bug tracking tool.

- ProblemTracker: It is from NetResults. It is fully web based collaboration software for bug tracking, defect tracking, issue tracking, and quality assurance.

- Bugzero: It is web based, easy to install, cross platform bug tracking system.

- Keystone: It keeps track of issues and tasks via a shared central resource, uses a SQL database and accesses the data via a web browser. Free for 1-9 active users. Commercial license required for 10-100+ active users. It is available From Stonekeep Consulting. Supports unix.

- RT (Request Tracker) is a trouble ticketing system that supports web, email or command line interfaces. database is based SQL This is written in perl and supports unix.

- wreq is a distributed request/problem tracking system with builtin knowledge database to help systems people handle requests and share knowledge among local support groups. This is written in perl and sports unix.

- ADT web: Advanced defect tracking web Edition is 100% web based tool for small. medium and large companies. It allows for tracking bugs, defects, suggestions by version, customer etc. Allows thousands of users to log in simultaneously from any location by use of their web browser.

2.10   Memory modelling

Functional simulation models are required for simulating with the memories used in the design. You can build your own models, download a free one or buy a commercial available memory model.

- Building own memory model: If there are enough resources available, you can spend time on building the memory model and use it for simulation. These can be build in any high level verification language or HDLs.

- Download freemodel: Memory vendors also provide free HDL models to download from. They might be highly accurate ones and available in Verilog or VHDLs from memory vendors for example Micron and Intel provide these.

- Buy commercial available models: Companies like Denali work with memory vendors to provide 100% accurate memory models for the components they manufacture. Good thing about commercial models is that since these companies are working with vendors before component is releases, you can get hold of simulation model as soon as memory is released and these are highly accurate models and gone through several simulations and testing already.

For details please refer to memory modeling chapter.

2.11   Regression

Regression running is essential for any verification process. There are tools and platforms available to help run thousands of tests simultaneously. You must choose appropriate tools to be able to run automated, repeatable and traceable regression suites, to deal with resource management, disk space and metrics reporting tools like bug tracking etc. Scripting also plays an important role in how you set up the regression. Perl is common scripting language to setup the regression environment and for reporting. LSF was used for load sharing to distribute jobs to ensure maximum utilization of available CPUs. For more details please refer to Regression chapter.

2.12   Functional Coverage

Functional Coverage is the determination of how much functionality of a design has been exercised by a verification environment. It requires the development of a list of functionality to be checked, the collection of data that shows the functionality of concern being exercised, and the analysis of the collected data. Functional coverage doesn't prove that the design properly executed the function, just that it was exercised. It is the job of the testbench to prove proper execution.

Traditional coverage techniques are not up to the task of determining the functionality exercised by a constrained, random test environment. Traditional techniques have relied on directed tests. If the tests passed, the functionality was exercised. With constrained random tests, data is randomly generated making it difficult to know specifically what will be tested beforehand. New techniques have been created.

2.12.1   Requirements for a Functional Coverage Tools

A functional coverage tool must have a certain set of requirements to be effective. This minimum set and their reasons are:

- Easy capture of coverage data.
- Easy creation of coverage metrics.
- Creation of coverage metrics should be reusable across multiple designs of similar types.

- Easy analysis of coverage data.
- Ability to accumulate results across multiple simulation runs.
- Ability to get snapshot of results during regression runs.
- Creation of reports.
- GUI.
- Ability to access the functional coverage tool during simulation and in a post-processing fashion.

2.12.2 Limitations of Functional Coverage

There are several limitations of functional coverage that should be taken into account. Some of them are:

- Knowledge of what functionality to check for is limited by the specification and the knowledge of the design.
- There is not any defined list of what 100% of the functionality of the design is and therefore there may be functionality that is missing from the list.
- It may take too long to create a coverage model to justify the investment.
- There is no real way to verify that a coverage model is correct except to manual check it.
- The coverage goals are not always easy to decide upon if the goal needs to be greater then one.
- The effectiveness of functional coverage is limited to the knowledge of the implementor.

2.13   Code Coverage

Code coverage is one of the quantifiable verification metrics to find out if the design has been completely verified. It is essential to incorporate some code coverage tool in verification processes. These tools provide various types of coverage metrics:

- Statement coverage: Also called line, block or segment coverage. Determines how many times each statement is executed.
- Toggle Coverage: Shows which signal bits in the design have toggled. This is used for both RTL and gate level netlists. Generally for gate level design fault coverage and power analysis.

- State machine coverage: Shows how many transitions of FSM were processed. This ensures that all legal states of FSM are visited and all state transitions are exercised.
- Visited State coverage: How many states of FSM were entered or visited during simulation. It is essential for complex state machines to find out if there are some unvisited states.
- Event Coverage: Also, called triggering. Shows whether each process has been uniquely triggered by each signal in its sensitivity list.
- Branch coverage: Also called decision coverage. Shown which case or "if..else" branches were executed.
- Expression Coverage: Also called condition coverage. Shows how well boolean expression in and if conditions or assignment has been tested.
- Path Coverage: Shows which routes through "if--else"and case constructs have been tested.
- Signal Coverage: Shows how well state signals or ROM addresses have been tested.

Features

- Has good performance and accuracy.
- Support distributed simulation and design environments.
- Should be able to determine untested code in the design.
- Should be able to reuse coverage metrics from other modules or IPs.
- Should be able to determine effective and efficient regression suite.
- Support automatic or manual filtering of uncoverable expression conditions.
- Support automatic finite state machine extraction.
- Single and multiple interacting FSM sequence coverage.
- Integrate easily into design flow and support various popular simulators and platforms.
- No changes to source code should be required.
- Extensive reporting that can be used for design reviews or code inspections.

Some Available Tools

- SureCov: Code coverage for Verilog. Supports automatic FSM extraction and analysis, comprehensive event and expression coverage, toggle coverage, easily integrated. Can be used for test suite ranking and merging too.

- HDLScore: Code coverage for Verilog, VHDL or mixed verilog/VHDL designs. Determines coverage for block, paths, expressions, variables, gate, FSM States and sequences, and toggle coverage.

- VN-Cover: Works with BHDL, Verilog or mixed language designs. Supports automatic FSM extraction, analysis and FSM path coverage. Provides coverage metrics for statement, branch, condition, path, triggering, trace, toggle, circuit activity analysis, FSM, State Arc, and Path coverage.

- Covermeter: Verilog code coverage tool that provides comprehensive coverage with features such as testbench grading, automatic extraction of FSMs, toggle coverage, source-line coverage, condition coverage, expression coverage, good performance and GUI.

CHAPTER 3

Using SCV for Verification

Verification takes a large portion of the overall design cycles. With complexities rising and schedules decreasing more productive means of verification are essential. In October of 2002 the OSCI standards body adopted SystemC Verification Extensions (SCV) to improve verification productivity. This chapter will discuss what is needed for verification, why the facilities of SCV were picked, and how SCV can be used in your designs to improve productivity.

After reading this chapter you should be able to:

- Understand the capabilities that are needed for a verification language.

- Get an introduction to SCV.

3.1   Features a verification language

A verification language needs to have the following capabilities to truly improve productivity:

- random data generation

- concurrently and synchronization

- data structures that can be used for checking.

These are the areas that are time consuming to create and maintain:

- Creation of real world stimulus - Directed random testing addresses this stimulus without exhaustively going through every possible combination of inputs.

- Need to drive the multiple inputs found on chips today - Concurrency and synchronization allows multiple inputs to be driven independently, as occurs in the real world, while still having some way to synchronize and control these independent interfaces.

- Need to efficiently check simulation results - Data Structures are used to quickly create protocols stacks and to create checking structures which are used by self-checking testbenches to determine if the device behaved properly.

Random data generation is necessary to cover the wide range of stimulus that can be seen by a hardware device today. Take an ethernet chip, for example. It needs to operate on 48 bit addresses. To truly prove that the chip can handle any legal address the verification engineer should try every combination. This is not practical. The traditional alternative is to execute simulations with values that are at the corners of the range, such as a low value, a value at the middle of the range, and a value at the high end of the range. This type of testing can miss combinations that do have problems. To hit every combination without exhaustive testing we turn to statistics. It has been proven that if a sufficient number of random values are chosen, the outcome of simulations of those values will be within a small percentage (less then 1%) of the outcome of simulations with all the possible combinations.

Concurrency and synchronization allow the creation and coordination of activity on multiple interfaces of a device. Many hardware devices these days have multiple interfaces, such as a PCI port, USB port, and an ethernet port. It is no longer acceptable to simply test the functionality of each port by itself. The verification engineer must be able to prove there are no hidden problems with the interaction of all of these ports. The only way this can be done is concurrently driving activity at all or a combination of the interface ports of a device. A verification language must have capabilities to control the creation of these activities and coordination. The SCV extensions relies on some of the underlying SystemC capabilities to create threads. A thread of activity is started by using the SC_THREAD macro. SCV then adds coordination facilities such as mutexs, semaphores, and mailboxes.

Data Structures are needed for stimulus generation and checking. All interfaces today require multiple data fields to be driven. For example, an ethernet packet requires a type, a length, a source address, a destination address, etc. It is much easier to create a software "struct" to capture and manipulate these fields then to create signals in an HDL and try to stretch the language to manipulate the signals. For checking, items like associative arrays and sparse memory arrays are useful to store and retrieve expected results from so that outputs can be checked.

3.2   Why use C++ for SCV?

In the "Overview of the Open SystemC" paper written in 1999, the initial section entitled "Motivation" stated: " Because C and C++ are the dominant languages used by chip architects, systems engineers and software engineers today, initiative members believe that a C-based approach to hardware modeling is necessary. This will enable co-design, providing a more natural solution to partitioning functionality between hardware and software". In a paper written by Synopsys in 1999 entitled "A New Standard for System Level Design" by Karen Bartleson, it stated "Momentum is building behind C/C++ class libraries as a solution for representing functionality, software and hardware at the systems level. The reason is clear: design complexity demands very fast executable specifications to validate the system concept, and only C/C++ delivers adequate performance. But without a common modeling platform, the market for system-level design tools will fragment with various proprietary languages and translators competing for attention, dramatically slowing technological innovation and economic progress."

The comments from the mentioned papers show that C/C++ is a good way to facilate a connection from system engineers and software engineers to the hardware designer. Since SystemC is simply a C++ class library it is natural to develop a verification language in C++ as well.

Many years ago when hardware languages were just beginning C/C++ wasn't a good choice for hardware design or hardware verification. That is because they didn't have basic hardware facilities such as time or signals. Therefore custom languages like verilog and VHDL were created to fill the gap. HDL's have matured to the point that it wouldn't make sense to replace them by C/C++. Hardware Verification languages (HVL's) are another story. The first HVL's came on the market approximately 8 years ago. While they are very useful for verification, they are focused at the hardware designer. As chips begin their migrations towards SOC and as their software content increases, system level verification and hardware/software verification needs must be considered in addition to the verification needs at the implemenation/HDL level. Since the industry is starting to adopt SystemC for system level modeling , it is natural for the verification to use be an extension to SystemC so that it can connect both worlds. This is what SCV does. It allows verification engineers to create an environment that can be used to drive both abstract system-level models and implementaiton models.

The use of SCV modeling requires knowledge of C++, which it is based upon. While SCV allows for a more seamless connection to System Level models and software environments, use of it needs to take into account the pre-requisite C/C++ knowledge.

3.3 What is randomization

Randomization in the verification space refers to the random creation of stimulus that drives a design. Randomization is often used in logic verification to generate unique data patterns that may be difficult to generate manually. These random data patterns help uncover bugs that may otherwise go undetected.

There are three types of randomization typically used in verification:

- Unconstrained randomization: data objects have an equal probability of taking on any legal value allowed by the data type of the object (e.g. an integer object can take on values between MIN_INT and MAX_INT). This can be done using any standard randomization utility; however, a major issue with this is the seed management of independent generator. SCV handles seed management for unconstrained randomization.

- Constrained randomization: the set of legal values that a data object can take is constrained by some algorithm. The constraint may be a simple limit of the legal values of the given data type, or it may be a complex expression involving multiple variables that are also being constrained.

- Weighted randomization: the set of legal values is weighted so that the probability of taking specific values is not uniform. For example, there may be a 60% probability that and integer object will take a value between 0 and 10, and a 40% probability that the value will be between 11 and 15.

Randomization can be used to choose types of tests to execute as well as to choose the data and control values that will be associated with each test.

There are six keys to a good constrained randomization solution:

- "The uniformity of distribution over legal values. For example, if the legal values to satisfy a constraint are {0, 1, 4, 5} each of these numbers must have a 25% probability of being selected.

- "Performance of constraint solver as the number of dependent variables grows. The number of independent variables plays a major role in the performance of any constraint solver. For a constraint solver to be useful, it must be able to handle a large number of independent variables.

- "Commutativity (order independence) of constraint equations. For example, the equations a = b must produce the same results as b = a.

- "The ability of the user to create relevant constraints involving multiple variables. This implies that the API for creating constraint expressions is rich enough to express complex constraints of typical systems.

- "Ability to debug over-constraints. When a constraint is unsolvable, it is critical that a user have a mechanism for figuring out why a constraint can't be met (e.g. is the constraint expression in error, or are unrandomized variables taking on invalid values).

- "The ability to control the value generation of constrained objects. This involves the ability to enable/disable randomization, and the ability to control pre-generation and post-generation work.

The SCV constrained randomization solution meets all of these requirements.

3.4   Introduction to SCV

3.4.1   Basic Purpose of SCV

The SystemC Verification extensions were created to increase the productivity of verification engineers. Verification engineers need many of the capabilities found in a pure software environment, such as pointers, abstract data structures, and object-oriented programming, in order to kept pace with design schedules.

With the inclusion of many different interfaces into a design along with the increase in complexity of those interfaces, verification is becoming more of a software oriented task then a hardware oriented task. Consider creating a verification model of a PCI-Express device. This model will need to include the layers defined for PCI-Express, which are the transport layer, the data-link layer, and the physical layer. Along with the transformation performed at each layer, the models must be able to inject and detect errors. Finally the specification is still evolving so that any work done creating this model may need to change several times. Creating this model from an HDL or hardware standpoint involves creating data structures like queues using languages that think in terms of real hardware. Therefore queues of a finite depth must be created. If software oriented languages are used then dynamically allocating memory is very easy and efficient. This eliminates the need to specific queue depths. If object-oriented techniques are used, then core functionality can be created in a base structure or class and specialized functionality can inherit from the base class and add the specialized capabilities. This way, when the specification changes only small portions of the verification model needs to change. In contrast, there can be a major re-write of the model if changes to the basic functionality are made.

What does the SCV extensions add to the Hardware Verification languages that are already in existence? There are several tools on the market that address the verification needs and several more are being proposed. All of these tools are focused on the verification of the digital content of a design. The SCV extensions allows verification work to start as early as system level definition and continue through the development of the digital hardware. As such it can provide a strong link between the different stages of a designs development. It becomes a verification contract between the Systems engineers, the Embedded Software engineers, and the Digital engineers. While there are chips in development today that don't require this link, I believe that it will become a necessity for any future SOC development.

3.4.2 Basic Language Constructs of SCV

There are four basic capabilities of the SCV extensions that were added for verification. They are randomization, control for concurrency, data structures, and transaction recording. In this section we will give the usages for each area, how to use the basic constructs, and an example.

Randomization is the first capability that you will probably use. As mentioned in the a previous section of this chapter, randomization is primarily used for stimulus generation. SCV allows its user the ability to create random numbers that are constrained by the type of data used. It allows for addition simple constraints with the keep_only/keep_out methods. It also includes a complex constraint solver, which allows relationships between random variables, and hierarchical constraints.

3.4.2.1 SCV Randomization

Scv_smart_ptr<data_type> name;   // Create a random variable

Name.keep_only (high value, low value); // Create a simple range constraint

Name.keep_out (high value, low value);   // Create a simple keep out constraint

This is the easiest way to create a random variable using SCV. You can create a random variable of any data type. The value will be initially constrained by the data type, when randomized. For example, a four bit integer will be constrained to be between 0 and 15. The keep_only and keep_out methods constrain the random values further. Note, multiple keep_only and keep_out calls can be made for the same variable. They will all be used when solving for a value. The "next" method is used to tell SCV to pick a new random value the next time the variable is used. An example program is:

```cpp
#include "scv.h"
// Declaration of top level testbench
SC_MODULE (sc_top) {
   void test();
   SC_CTOR (sc_top) {
      SC_THREAD (test);
   }
};
// Test to execute
void sc_top::test () {

scv_smart_ptr<uint> addr; // Declare a random unsigned integer
scv_smart_ptr<uint> data; // Declare a random unsigned integer

// Pick a random value for addr & data from the natural
// integer range. Note, the * is needed to get the value of the smart pointer.
cout << "Addr: " << *addr << " Data: " << *data << endl;

// Set simple constraints for the random variables
addr.keep_only (10, 100); // Constrain addr to the range 10 - 100
addr.keep_out (20, 30);    // Constrain addr to avoid the range 20 - 30
addr.keep_out (75, 90);    // Constrain addr to avoid the range 75 - 90
                    // The Addr value can only be picked from 10-19, 30-74, 91-100;
data.keep_only (10, 100); // Constrain data to the range 10 - 100

// Pick several random values for addr and data
for (int i=0; i < 5; i++) {
   // The next method tells SCV to pick a new random
   // value next time the variable is used.
   addr.next();
   data.next();
   cout << "Addr: " << *addr << " Data: " << *data << endl; // Print out the values of the variables.
}
}
```

3.4.2.2 Concurrency

Concurrency is the ability to have different verification activities executing at the same time. This is accomplished in SystemC with SC_THREAD. Every time a module starts the user has the ability to start a thread that is separate from the simulator. In the example given in 3.4.2.1 the test method was started in a separate thread:

```
#include "scv.h"
// Declare a SystemC module named sc_top
SC_MODULE (sc_top) {
    void test();
    SC_CTOR (sc_top) {
        // The SC_THREAD call will start executing the test method in a separate thread
        SC_THREAD (test);
    }
};
```

Concurrency is not unique to a verification language. All HDL design languages such as Verilog or VHDL are built upon the notion of concurrency. Every module or entity is by definition running concurrently. So what additional advantage does a verification language bring? The additional advantage is the ability to synchronize across the numerous concurrent verification activities that may be executing.

In the HDL languages, the only way to synchronize concurrent activity is through the use of signals. Normally these signals need to be included as part of the port list of the module that will execute.

If the device you were testing had several ports and there was a need to drive all ports with stimulus at the same time, you could start it as follows:

```
#include "scv.h"
// Declare a SystemC module named sc_top
SC_MODULE (sc_top) {
    duv dual_port_ram;              // Instaniate the DUV and name it dual_port_ram
    sc_signal<sc_unit<16> > addr0;   // Create signals addr0, data0, addr1, data1, read0, read1
    sc_signal<sc_uint<16> > data0;
    sc_signal<sc_uint<16> > addr1;
    sc_signal<sc_unit<16> > data1;
    sc_signal<bool> read0;
    sc_signal<bool> read1;
    void port0_test() {             // This is the test that will execute for port0
        cout<< "execute port0 test"; )
        // Do several reads and writes on port0
    }
    void port1_test() {             // This is the test that will execute for port0
```

```
        cout<< "execute port1 test";
        // Do several reads and writes on port1
        ....
    }
    SC_CTOR (sc_top) {
      // Create a seperate thread for each port test; Both tests will execute concurrently
      SC_THREAD (port0_test);
      SC_THREAD (port1_test);
    }
};
```

3.4.3 Examples of an SCV testbench

Lets review an example of an SCV testbench. The example shows how you can
generate a constrained set of cpu instructions with a small amount of code.

```
void test_code::test () {

  // Create a test that will randomly
  // Initialize
  list<cpu_opcode_type> opcode_list;
  list<register_type> operand1_list;
  list<register_type> operand2_list;

  // Add constraints to list
  opcode_list.push_back(ADD);
  opcode_list.push_back(ADDI);
  opcode_list.push_back(ANDI);
  opcode_list.push_back(XOR);
  operand1_list.push_back(REG0);
  operand2_list.push_back(REG2);// Set constraints
  instruction.inst->opcode.keep_only(opcode_list);
  instruction.inst->operand1.keep_only(operand1_list);
  instruction.reg->keep_only(operand2_list);
  instruction.byte->keep_only(0x5);

  // Execute
  for (uint i = 0; i < 7; i++) {
    generate_instr (instruction);
    wait (clk.posedge_event());
  }
  sc_stop();
}
```

3.5   Creating A Stimulus Generator using SCV

The focus in this section will be the recommended structure of transactors and test-benches, the layering of random data that will be driving the various interfaces, and the use of synchronization mechanisms to control the interaction of the various interfaces.

3.5.1   Testbench Structure

Testbenches normally contain the DUT (Device Under Test), verification intellectual property (IP) components that model external and internal interfaces (such as RAMs), and verification elements needed for self-checking purposes. The SCV extensions adds an additional dimension to these components. Since the SCV extensions can drive both a system-level model based on SystemC and an HDL model, the layout of the testbench needs to allow reuse of common components across both these domains.

A typical testbench will look like:

```
// Declare a SystemC Module name top
SC_MODULE (top) {
    duv duv1;    // Instantiate the DUV
    test test1;    // Instantiate the test
    transactor trans1;    // Instantiate a transactor
    transactor trans2;    // Instantiate another transactor
    sc_signal<sig_type> sig_name: // Create a signal of type sig_type and use it to make connections

    ....
SC_CTOR (top) :
    duv1 ("duv1")
    , test1 ("test1")
    , trans1 ("trans1")
    , trans2 ("trans2")
    , sig_name ("sig_name")

    .....

    {
    // Connect the testbench
    duv.sig1 (.....);          // Connect a signal of the duv to a signal in the transactor

    ....
    trans1.sig(...);          // Connect a signal of the transactor to a signal in the design

    ....
    trans2.sig(....);          // Connect a signal of the transactor to a signal in the design

    .....
    test1.port1(trans1);    // Connect port1 to the first transactor
    test2.port2(trans2);    // Connect port2 to the second transactor
    }
    }
```

The test routine is executed once the testbench starts, which starts the simulation.

3.5.1.1 Test

A test typically controls all aspects of the simulation. This includes driving transactors with random data, synchronizing the execution of different transactors in a testbench, collecting results information, and checking results. While all this functionality can be contained in a test, it makes the test less reusable. That is because the test must contain some knowledge of the DUT in order for checking to be performed. If the DUT changes, the test must also change.

The test should really be only a stimulus generator. The checking should be done by a verification item called a Response Checker. The job of the Response Checker is to verify the results of the simulation. This will be discussed more in the chapter 5 on testbench components. The test should instantiate various transactors and then call the transactor routines.

```
// Declare a SystemC module named test
SC_MODULE (test) {
sc_port <transactor_interface> transactor_port;  // This port will connect to a transactor
void run ();
SC_CTOR (test) {
   SC_THREAD (run);  // Start a thread named run
}
// Define the run routine
void run() {
   data_structure data;   // data_structure is a struct
   for (int i = 0; i <= 1000; i++) {
     transactor_port.write(data); // randomly populate data structure and perform write to duv
     transactor_port.read(data); // perform read from duv
   }
}
```

3.5.1.2 Transactor Structure

Transactors drive the stimulus into a design. An SCV transactor should be created such that it can drive a TLM (Transaction Level Model), which is typically used for System Level designs, or an RTL model, which is used for the implementation of a design. The structure of a SCV transactor should contain:

- Ports
- Tasks
- Initialization Routines

- Verification Infrastructure

The ports of an SCV transactor can be a set of signals or a SystemC channel. The set of signals will allow the transactor to drive an RTL and may appear easier to use at first. These type of ports look like:

sc_in <data_type> signal_name;

The disadvantage to this type of port is that it is limited to only the RTL domain. If you create a port as a channel, the channel can either connect to an RTL model or a TLM.

sc_port<interface> port_name;

3.5.2 Randomization

Section 3.3 takes a look at the different forms of randomization. In this section we will focus on an example using constrained randomization in SCV. But first, let us take a look at some definitions related to randomization in SCV, and then the actual mechanics of how these are implemented.

There are basically two ways to specify constraints in SCV. Using:

Simple Constraints:

When a variable needs to be constrained but it does not have constraint relationships with any other variables, this is usually declared using simple constraints. An examples of simple constraints in SCV are the keep_only/keep_out constructs, and the set_mode(scv_bag<T> / SCAN /RANDOM_AVOID_DUPLICATE) constructs. A simple constraint in SCV is defined as basically any construct that constrains the value of a single variable and which does not use the SCV_CONSTRAINT, or SCV_SOFT_CONSTRAINT constructs. Simple constraint constructs can be used both within constraint classes and outside of constraint classes.

Complex Constraints:

Complex constraints are used when the variable to be constrained has relationships with other variable. They are often specified using C++ expressions that evaluate to a simple Boolean value. They are usually constraint expressions that appears within the SCV_CONSTRAINT/SCV_SOFT_CONSTRAINT constructs. Complex constraints can only be specified within constraint classes and appear within the constructors of constraint classes. An example of a complex constraint expression is the following:

SCV_CONSTRAINT(a() > b() && b() > c() && (a() − c() > 100));

The complex and simple constraints are grouped together using constraint classes, so that they can be used to generate constrained random stimulus. A constraint class is any class that derives from class scv_constraint_base.

The SCV constraint solver generates new values for output variables which are defined using either SCV_CONSTRAINT, or SCV_SOFT_CONSTRAINT

Example of simple constraints:

```
// Create a structure will a Addr field of type int and a DATA field of type int
// Constrain ADDR to a range of 0 - F0FF but keep out of the range A012 - A0FF
// Constrain DATA to a range of 001C - 0FFF but keep out of the range 0A00 - 0C88
struct cnstr_simple: public scv_constraint_base {    // scv_constraint_base is a built in SCV class
    scv_smart_ptr <int> ADDR;                         // scv_smart_ptr's are used to create random vars
    scv_smart_ptr <int> DATA;
    SCV_CONSTRAINT_CTOR(cnstr_simple) {       // Default constraints are put in constructor
        ADDR->keep_only(0x0000, 0xF0FF);
        ADDR->keep_out(0xA012, 0xA0FF);
        DATA->keep_only(0x001C, 0x0FFF);
        DATA->keep_out(0x0A00, 0x0C88);
    }
};
```

Example of complex constraint:

```
// Create a structure with an ADDR field of type int and a DATA field of type int
// Constrain ADDR and DATA with a relationship between them. This is called a complex constraint.
//     If (0x00FF < ADDR < 0xFFF) and (DATA > 0xF55) then (ADDR + DATA <= MAX_MEM_INDEX)
struct constr_complex: public scv_constraint_base {
    scv_smart_ptr <int> ADDR;
    scv_smart_ptr <int> DATA;
    SCV_CONSTRAINT_CTOR(constr_complex) {
        SCV_CONSTRAINT(
            (ADDR() >= 0x00FF) &&  (ADDR() <= 0x0FFF) &&
            ( DATA() >= 0x0F55) &&
            ( ADDR() + DATA() <= MAX_MEM_INDEX)
        );
    }
};
```

Before we dive into an example showing hierarchical constraints, let us take a quick look at how random variables are defined using smart-pointers. Smart pointers are objects in C++ that mimic the capabilities of normal "dumb" C/C++ pointers, but which also have intelligent capabilities added such as automatic memory management, garbage collection and randomization.

Example:

```
struct packet_t {
    sc_uint<32> X;
    sc_uint<32> Y;
};

struct packet_base_constraint : public scv_constraint_base {
    //create a packet object
    scv_smart_ptr<packet_t> packet;
        ....
}

void test::test1 () {
    packet_base_constraint p;

    for (int i=0; i < 10 ; i++) {
        p.next();
        cout << "packet values for p = " << p.print() << endl;
    }
}
```

The smart pointer in the above example will create random numbers for the X and Y fields in the packet_t structure every time p is used after a p.nex() call. The p.next() tells the software to randomize the smart pointer values the next time they are used. Note, the ability to randomize entire structures, instead of calling the next() method for every item in the structure. Also note that when the test completes, the garbage collection for the smart pointer takes place automatically.

Let's now look at a simple example where four different opcodes [ADD, DIVIDE, SUB, INC] are generated randomly corresponding to two different types of instructions [Register, Immediate] also generated randomly. The SUB & INC opcodes are type immediate, and DIVIDE & ADD are of type Register. Let us look at a couple of different ways the generation of the opcodes can be constrained.

```
#include "scv.h"
```

```cpp
// General Information:
//
// The goal of this code is to be to test a CPU which has two kinds of
// instructions, immediate opcodes and register opcodes. The test gives the
// the capability of defining two dimensions of randomization:
//        . Set a weighted distribution for selecting an individual opcode
//        . Set a weghted distribution for selecting an opcode type (register, immediate)
// Note the inter-dependence of these dimensions. This would be considered a hierarchical
// constraint because the higher level, opcode type, will affect the selection of the lower level,
// the actual opcode.
//
// The opcodes and opcode types are:
//       opcodes: add, divide, sub, inc
//       opcode types:   reg opcodes  = add, divide
//                       imm opcodes = sub, inc
//
// The test structure allows the randomization constraints to be specified, for example, as:
//        - 70% of the time pick an opcode type of imm.
//           .Of the imm opcodes, 66% of the time pick add and 33% of the time pick divide
//        - 30% of the time pick an opcode type of reg.
//           . Of the reg opcodes, 80% of the time pick inc and 20% of the time pick sub
// The main pieces of this code are:
//    . The instructions,
//        - These are defined in the enum opcodes and they are add, divide, sub, and inc.
//    . class opcode_gen
//        - This structure contains all of the pieces necessary to set, reset, and
//          create the randomization.
//      . SC_MODULE(test)
//        - This is the actual test.
//        - It drives two scenarios.
//           . Scenarion 1
//              - Weighting on individual opcodes: 20 Add, 10 Divide, 5 Sub, 20 Inc
//              - Weighting on opcode type: 30% reg, 70% immed
//           . Scenario 2
//              - Weighting on individual opcodes: 20 Add, 11 Divide, 1Sub, 9 Inc
//              - Weighting on opcode type: 75% reg, 25% immed


//***********************************************************************
// This code defines all the basic data types needed for the test
//***********************************************************************
//
// This defines the opcodes to be used by the test
enum opcodes { add, divide, sub, inc };
//
// This code defines how to print the opcode information to the screen.
ostream& operator<< (ostream& os, const opcodes& oc)
{
  switch(oc)
```

```cpp
  {
    case add: os << "ADD"; break;
    case divide: os << "DIVIDE"; break;
    case sub: os << "SUB"; break;
    case inc: os << "INC"; break;
    default: os << "<UNKNOWN>"; break;
  }
  return os;
}
//
// This code defines a special SCV class that allows an ordinary data type or structure,
// which in this case is "enum opcodes", to have randomization capability
template<>
class scv_extensions<opcodes > : public scv_enum_base<opcodes > {
public:
  SCV_ENUM_CTOR(opcodes) {
    SCV_ENUM(add);
    SCV_ENUM(divide);
    SCV_ENUM(sub);
    SCV_ENUM(inc);
  }
};


//*************************************************************************
// This code defines the class opcode_gen, which collects all the data types
// and routines that provide the two-dimensional randomization of opcodes
// that was defined in the General Information section at the top of this code.
//*************************************************************************
class opcode_gen
{
  public:

    //the distributions that are set for immediate and register codes
    scv_bag<opcodes> reg_codes, imm_codes;

    //the distribution set up to chose a code type to generate
    scv_bag<scv_smart_ptr<opcodes> > code_types;

    //the actual value generators
    scv_smart_ptr<opcodes> reg_gen, imm_gen;

    //variables for to simplify the user interface
    bool isset, distset;
    int  reg_weight, imm_weight;

    // Constructor for class opcode_gen; The constructor initializes the class
    opcode_gen() :
      isset(false), distset(false), reg_weight(0), imm_weight(0)
    {
```

```
  }
  // The next method prepares all the elements of the class to select a new random value the
  // next time the class is asked to produce values. The next method is built into SCV and
  // therefore doesn't normally need to be defined. In this case their is a need to enhance
  // the next method with additional code.
  opcodes next()
  {
    if(!distset)
    {
      reg_gen->set_mode(reg_codes);
      imm_gen->set_mode(imm_codes);
      code_types.clear();
      if(reg_weight) code_types.add(reg_gen, reg_weight);
      if(imm_weight) code_types.add(imm_gen, imm_weight);
      scv_smart_ptr<opcodes> type_gen_p = code_types.peek_random();
      type_gen_p->next();
      return type_gen_p.read();
    }
  }


  // The set_type_weight is a supporting routine that makes it easier for the test writer to
  // set the weighted distribution for the opcode types. The test writer would need to explictly
  // set the individual values (reg_weight, imm_weight, distset) if this routine wasn't defined.
  void set_type_weight( bool reg_type, int weight )
  {
    if(reg_type) reg_weight = weight;
    else imm_weight = weight;
    distset = false;
  }


  // The set_weight is a supporting routine that makes it easier for the test writer to
  // set the weighted distribution for the opcode types. The test writer would need to explictly
  // set the individual weight values if this routine wasn't defined.
  void set_weight( opcodes val, int weight )
  {
    switch(val)
    {
      case add:
      case divide:
        reg_codes.add(val, weight);
        break;
      case sub:
      case inc:
        imm_codes.add(val, weight);
        break;
    }
    isset = true;
  }
};
```

```cpp
//*********************************************************************
// This is the test code
//*********************************************************************

// The top level module is called test
SC_MODULE(test) {
  opcode_gen gen;
  _scv_associative_array<opcodes, int> c_cnt;

  // This is the constructor for the top level module named test. The constuctor
  // initializes all of the variables in this module and starts up a new thread,
  // ie SC_THREAD (doit), which will execute the test method which is named doit
  SC_CTOR(test) : c_cnt("code_count", 0) { SC_THREAD(doit); }

  // The doit method is the actual test
  void doit()
  {
    //Create distribution Constraint-Set#1 which constrains the test as follows:
    //       - 70% of the time pick an opcode type of imm. (gen.imm_weight = 70)
    //             .Of the imm opcodes, 66% of the time pick add and 34% of the time pick divide
    //               (gen.reg_codes.add(add, 20);gen.reg_codes.add(divide, 10);)
    //               The percentages come from the number of elements (20 for add, 10 for divide),
    //               divided by the total number of 30 (20+10) elements. Therefore the percentage
    //               for add is 66% (20/30) and divide is 33% (10/30).
    //       - 30% of the time pick an opcode type of reg. (gen.reg_weight = 30)
    //             . Of the reg opcodes, 80% of the time pick inc and 20% of the time pick sub
    //               (gen.reg_codes.add(sub, 5);gen.reg_codes.add(inc, 20);)
    //               The percentages come from the number of elements (5 for sub, 20 for inc),
    //               divided by the total number of 25(5+20) elements. Therefore the percentage
    //               for subis 20% (5/25) and divide is 80% (20/25).

    gen.reg_codes.add(add, 20);
    gen.reg_codes.add(divide, 10);
    gen.imm_codes.add(sub, 5);
    gen.imm_codes.add(inc, 20);
    gen.reg_weight = 30;
    gen.imm_weight = 70;
    gen.isset = true;
    gen.distset = false;
    //
    opcodes oc;

    // Loop 101 time, each time selecting a new opcode
    for(int i=1; i<101; ++i)
    {
      oc = gen.next();
      c_cnt[oc]++;
      cout << oc << " ";
```

```
    if(i%10 == 0) cout << endl;
  }
  print_stats();

  //Create a new distribution Constraint-Set#2 which constrains the test as follows:
  //       - 25% of the time pick an opcode type of imm. (gen.imm_weight = 25)
  //            .Of the imm opcodes, 65% of the time pick add and 35% of the time pick divide
  //              (gen.reg_codes.add(divide, 1);)
  //            In this case, the test is reusing the previous definition of the number of elements of
  //              add (20) and divide (10) , and then adding one element for the opcode divide(10+1).
  //            The percentages come from the number of elements (20 for add, 11 for divide),
  //              divided by the total number of 31 (20+11) elements. Therefore the percentage
  //              for add is 66% (20/31) and divide is 33% (11/31).
  //       - 75% of the time pick an opcode type of reg. (gen.reg_weight = 75)
  //            . Of the reg opcodes, 10% of the time pick inc and 90% of the time pick sub
  //              (gen.reg_codes.add(sub, 1);gen.reg_codes.add(inc, 9);)
  //            The percentages come from the number of elements (1 for sub, 9 for inc),
  //              divided by the total number of 10(1+9) elements. Therefore the percentage
  //              for sub is 10% (1/10) and divide is 90% (9/10).

  gen.set_type_weight(IMM_TYPE, 25);
  gen.set_type_weight(REG_TYPE, 75);
  gen.set_weight(divide, 1);
  gen.set_weight(inc, 9);
  gen.set_weight(sub, 1);

  c_cnt.clear();  // Clear the count array so that we start counting for zero

  // Loop 101 time, each time selecting a new opcode
  for(int i=1; i<101; ++i)
  {
    oc = gen.next();
    c_cnt[oc]++;
    cout << oc << " ";
    if(i%10 == 0) cout << endl;
  }
  print_stats();

  sc_stop();  // Stop the simulation
}
void print_stats() {
  cout << "\n------------------------------\n";
  cout << "Constraint Stats: \n";
  cout << " Immediate opcodes generated = "
       << c_cnt[sub]+c_cnt[inc] << "\n"
       << " Register opcodes generated = "
       << c_cnt[divide]+c_cnt[add] << "\n"
       << "  SUB = " << c_cnt[sub] << "\n"
       << "  INC = " << c_cnt[inc] << "\n"
```

```
        << "  ADD = " << c_cnt[add] << "\n"
        << "  DIVIDE = " << c_cnt[divide] << "\n";
    cout << "------------------------------- \n \n" << endl;
  }
};
NCSC_MODULE_EXPORT(test);

Results from Constraint-set#1
    Immediate opcodes generated = 73
    Register opcodes generated = 27
        SUB =  17
        INC =  56
        ADD =  18
        DIVIDE =  9

Results form Constraint-ste#2:
    Immediate opcodes generated = 28
    Register opcodes generated = 72
        SUB =  4
        INC =  24
        ADD =  45
        DIVIDE =  27
```

3.6   Future of SCV

SCV was ratified by the OSCI standards body in October of 2002. It is now starting to be utilized on real projects. The future of SCV looks very bright. It has been available only a little over a year and yet there have been publications of success. A couple of examples of this are:

- One of the best paper awards at DVCon 2004 went to Starkey Labs for their paper on using SystemC and SCV for the design and verification of a custom DSP processor.

- AMD presented a paper describing their implementation of a functional coverage capability in SystemC, at the SystemC Users Group in Europe in February 2004.

There are a number of other similar examples. In general, the SCV extension to SystemC provides a complete, practical, and open source way to drive directed random verification for both abstract designs at a System Level and a detailed RTL Level design. This has many people excited about the possibilities.

CHAPTER 4

Functional Verification Testplan

The verification test plan gives a measurable view for thorough evaluation of design functionality. It acts like a scoreboard to help make right decisions and provides the guidance for management and technical efforts throughout the verification phase in design cycle. It establishes a comprehensive plan to communicate the nature and extent of testing that is necessary for the tapeout. This plan is used to coordinate the orderly scheduling of the events by specifying tools and resource requirements, the verification methodology to be employed, a list of tests stimulus to be applied, and schedule for regression running for meeting sign off criteria. Finally, it provides a written record of required input stimulus, specific criteria that each functionality in the spec must meet, execution instructions, and expected results of design functionality.

One of the major reasons for inadequate testing is inadequate testplan which poorly specifies requirements that are unclear, too general and have unrealistic schedule. A good testplan should have clear, complete, detailed, cohesive, attainable and testable requirements that are agreed by all. There should be adequate time for planning, design, testing, bug fixing, re-testing, changes, documentation and executing a rigorous test bench to build confidence on quality and functionality of the RTL.

Topics covered in this chapter:

- What is the best testplan with examples on how to implement.
- Testplan format.

- Purpose of testplan.
- Product phases.
- Test strategy.
- Test activities.
- Schedule.
- Resource requirement.
- Risks and concerns.
- Sign off criteria - verification metrics.

A complete verification plan should discuss everything from tools used, details of verification environment, methodology for automatic checking, regression running, bug filing and metrics for sign off. A test plan should be reviewed thoroughly by all the relevant experts of design.

First of all, the functionality to be tested should be broken down into more detailed list of features that are to be verified. This list of features gives an initial list of tests that are to be executed for verification. As the design is implemented in parallel to the verification and as the tests are run and bugs are found and various coverage metrics are gathered and analyzed, more tests need to be added to this test plan.

4.1   Different kinds of Tests

Testplan describes different tests that are required to verify the functionality. Tests can be described in wide variety of categories:

- Black box tests: System level testing is mostly black box testing. Tests are based on the requirement and functionality of the design, there is no knowledge of the internal design or code.
- White box tests: Block level tests are mostly white box tests based on internal logic of the RTL. Tests are based on coverage of code statements, branches, paths and conditions.
- Unit tests: These are to test particular function or code at micro scale. This is also done by block designers as it requires detailed knowledge of internal code and design.
- Incremental integration testing: continuous testing of a RTL as new functionality is added to the design. This requires that simulation at top level should be able to run using block plug and play mode.

- Integration testing: testing of combined parts of blocks to determine if they function together correctly. This is different than toplevel where all the system is tested together.

- Functional testing: black-box type testing geared to functional requirements of an application. These tests are done by verification engineers.

- System testing: black-box type testing that is based on overall requirements specifications, covers all combined parts of a system.

- End-to-end testing: similar to system testing; the 'macro' end of the test scale; involves testing of a complete application environment in a situation that mimics real-world use, for example running system simulation of traffic management chip with actual framer, cell/switch fabric models.

- Sanity testing: typically an initial testing effort to determine if a new RTL version is performing well enough to accept it for a major testing effort.

- Regression testing: re-running tests after fixes or modifications of the RTL or its environment. It is generally difficult to determine how much regression running is needed, especially near the end of the design cycle. Automated regression setup can be especially useful for this type of testing.

- Acceptance testing: final testing based on specifications of the customer, or based on use by end-users/customers over some limited period of time.

- Stress testing: term often used interchangeably with 'load' and 'performance' testing. Also used to describe such tests as system functional testing while under unusually heavy loads, heavy repetition of certain actions or inputs, input of large stimulus for chip capacity/buffer management.

- Performance testing: term often used interchangeably with 'stress' and 'load' testing. To determine system performance under stress or normal conditions and see if they match expected values.

- Recovery testing: testing how chip recovers after tearing down the connection or stopping the input generation entirely and restarting it.

- Exploratory testing: often taken to mean a creative, informal tests that is not based on formal test plans or test cases. Verification engineers may be understanding specs while performing these tests. This helps in making sure that functional specs are well understood and are clear.

- Comparison testing: comparing design weaknesses and strengths to competing products.

- Corner tests: Verification engineers work with designers to define additional tests that address the corner cases for optimum coverage.

- Negative testing: Deliberately inserting some errors in the test stimulus and check if RTL catches it and that system does not hang on seeing negative input that is out of specified range.

4.2   When to Start

Early in the design cycle when functional specs for chip are being written, the effort for generating a test plan also begins. Familiarity with the design and specialized verification knowledge is one of the main requirement to put together the test plan. Both RTL design and verification are parallel processes. First and foremost task of any verification effort is to create a testplan. One of the known truth of whole verification effort is that it can only detect presence of errors not their absence. So, it becomes more important to have an effective testplan. This can be divided into two documents

- Verification methodology document that mainly describes the tools, resources and methodology used for testing.
- Tests status document that describes the details of the tests and also maintains daily status and metrics of the tests.

A good verification plan accurately outlines different tasks that are required for completing functional verification for successful tapeout and a good verification methodology assures maximum testing within the limitations of time and resources as per schedule.

Flow is as defined in the figure 4-1. From Fspecs, chip definition, description of input/output interfaces, system environment, performance objectives and schedule requirements emerges first methodology document. After that it goes through various review cycles before actual detailed test document is formed along with the implementation of all the verification infrastructure. The initial plan for verification identifies and implement methods on how testing should be done. Status for the test plan coverage is tracked with each regression run until final sign off criteria is met.

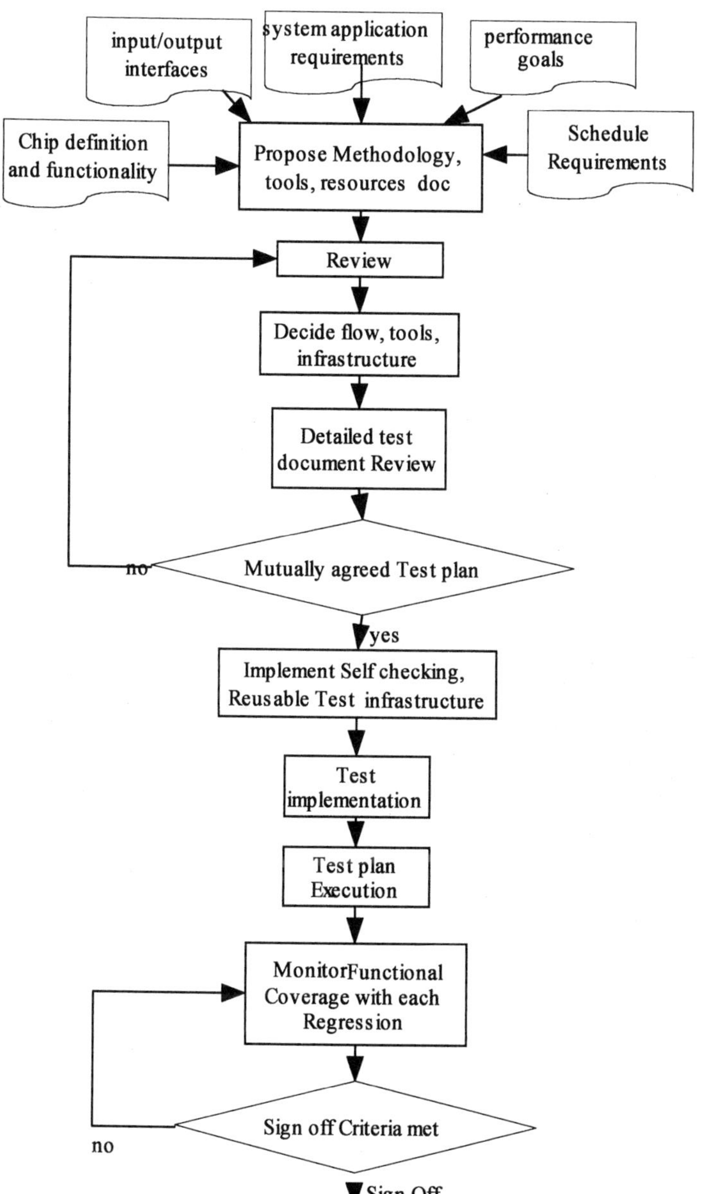

Figure 4-1. Verification Flow

4.3   Verification plan document

System level testplan describes the detailed methodology and approach used for testing. This is initial planning phase. To prepare an effective testplan you need to describe the purpose of testplan, process to review and methodology followed.

Purpose of testplan is to give an overview of what will be tested and what will not be tested. In following paragraphs this is described using an example of networking traffic management design under test (DUT). The sections for which detailed features can not be discussed, description of what is to be covered is provided.

4.4   Purpose and Overview

This document addresses how to verify the networking device for traffic management and SAR application, including its testbench functional elements and the test suite. It also covers resource requirements and dependencies. Another document called test status/description will evolve from this high level test plan document which will have detail test names and description to achieve all the functionality

4.4.1   Intended Use

This document is intended as a living document for the use of the engineering group. The contents of verification plan will evolve as new scenarios and further details on verification are flushed out as the project progresses.

4.4.2   Project References

References for testplan are

- Functional specification for the design
- Draft proposed for interface documents
- Verification methodology document
- Software interface specification
- Standard documents for standard interfaces for compliance checking

4.4.3   Goals

This networking chip implements a multi-service approach with dynamic bandwidth allocation among the various flows carrying the different types of traffic. It is designed such that it can be used in different applications that fit the needs of various system architectures. Two of the more popular applications are traffic manager and SAR solutions for OC-48 line cards in routing and switching systems. The goal of the plan is to make sure that the RTL performs the intended functionality of the specification, also known as functional verification. It is assumed that for gate level verification besides doing formal equivalence checks and static timing verification, simulation will have to be run. Metrics for the code and functional coverage are also described at the system level

This document focuses on functional verification at the system level. It is assumed that block level code coverage and verification will be covered in separate spec owned by block designers and they will certify blocks functionality against the specifications. Block verification can be accomplished by using block level tests and RTL checking tools like lint or assertion based tools along with the block test bench. The plan for block verification is considered outside the scope of the document.

This plan covers the various use models of the chip under verification in the line card used in conjunction with cell-based switching fabric and packet routers. The two main application modes for the device are discussed below.

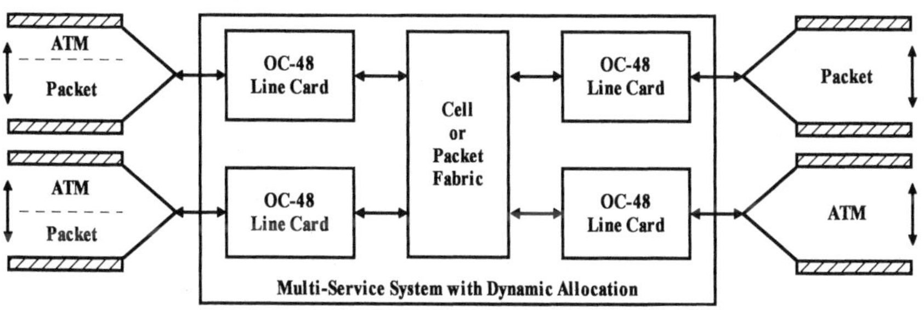

Figure 4-2. Multi-service system

Above figure 4-2 depicts an exemplary system level overview, illustrating how the device may enable OC-48 line cards in a router or switch system to receive different types of traffic, to dynamically allocate bandwidth among the various flows, to adapt the incoming traffic to the requirements of the cell-based or packet-based switch fabric, and to convert among the different traffic types for output.

In Figure 4-2, each line card can receive both ATM and packet traffic coming in on different flows across a fiber optic connection.   The device on each line card, acting as an ingress device interfacing with a framer device, can receive, buffer and adapt the incoming traffic to the switching requirements of either the cell or packet switch fabric. On the egress side, the traffic can be sent out to the framer device either in the same traffic type (e.g., ATM in, ATM out) or in a different traffic type (e.g., ATM in, packet out or packet in, ATM out). With regard to dynamic bandwidth allocation, the DUToffers extensive QoS capabilities to shape and schedule the flows as desired according to their QoS priorities.

4.4.4 System Level Testbench:

As described in the figure 4-3 below, the system level testbench provides the verification environment to test the devices. There are multiple interfaces to the chip, the SPI4 bus, PL3 bus and the CPU interface. Utopia, 2/3, PL2, CSIX, NPSI were also supported in later releases of the device. On the left side of the chip is the SPI4 input bus interface and PL3 (PHY and LINK) input interface bus communicating via the SPI4 input TVM or PL3 input TVMs (PHY or LINK) to the cell/packet generators. In addition there is the CPU interface to the chip, which is primarily used to setup the chip configuration. The generators are driven by a test stimulus, which comprise the commands and options to run the packet/cell generator. On the right hand side of the figure, there is the SPI-out bus, PL3 bus as well as the DUT to CPU bus. The SPI-out bus talks to a packet checker, which does low level checks such as bit level/field level checks. Similarly traffic can be sent in and expected at PL3 interface bus. There are three input interfaces to generate traffic (SPI, LINK and PHY) and tests randomly select one or multiple interfaces for input and output. The SPI4 or PL3 bus is monitored using the SPI4 or PL3 bus monitors, which checks for transaction level accuracy. They could optionally also log the traffic activity and error messages into the corresponding log files. On top of the bus-monitors, there are rate monitors, which monitor the rate of traffic. The input and output rate monitor finally interacts with a high-level response checker, which could determine the outcome of the test. These blocks are further described in detail below.

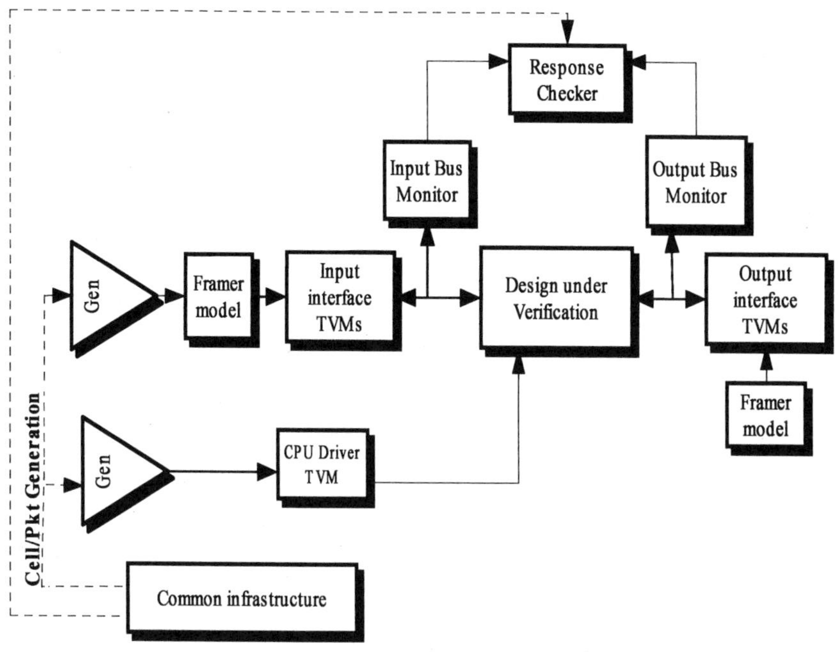

Figure 4-3. system level testbench

4.4.5 DUV

Design under verification comprises of chipset configured in various modes of operation which are described further during system testing.As shown in figure 4-4 the DUV will have various internal bus monitors to aid system debugging. These are present on the interfaces of the blocks to track the state of the packet/cell as it goes through the system. These monitors act as passive monitors and record the activity into the SST2 database. These monitors will provide following function:

-Track the cell/packet through the device at each interface.
-Log the information and pass it on to checkers.
-Flag any error against any interface conditions being checked on interface signals.
-Can flag EOP/SOP errors.
-Can flag Parity errors.
-CRC errors, bad packet.
-Report the rate of traffic for time window.

Basically, each monitor besides passing on the cell information to the checker will also pass on any error information to be acted upon by the central messaging task. Any other checks or info that central messaging task needs from the block interfaces will be provided by monitors

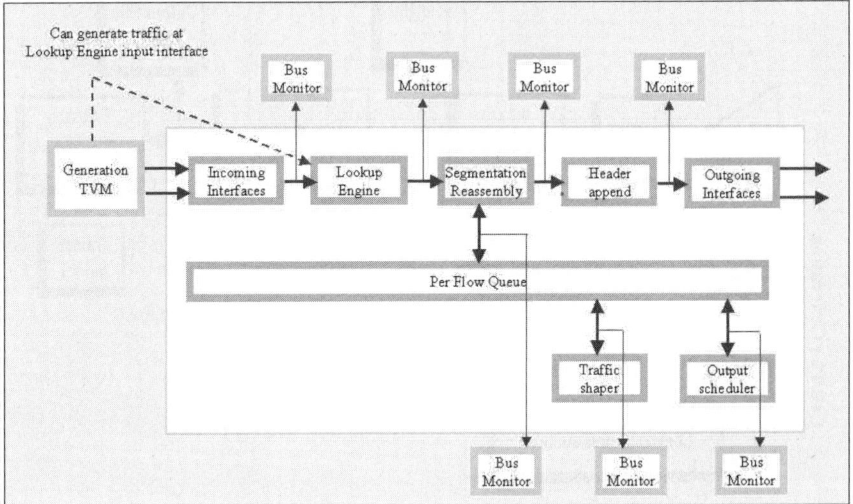

Figure 4-4. DUV with internal monitors

4.4.6 Transaction Verification Modules

There are multiple interfaces to the chip. SPI4, PL3, PL2, UTOPIA 3, UTOPIA 2, CSIX, NPSI serves as an interface to the framer(s) and the CPU interface serves as an interface to the CPU. The corresponding TVMs serve as an abstraction layer from the packet generation and checkers as well as the CPU to the chip setup commands to the signal level protocol implementation. Therefore the packet generation and packet checkers along with the CPU command generation serve as the main interface to the test stimulus. The general architecture for TVMs is shown in figure 4-5

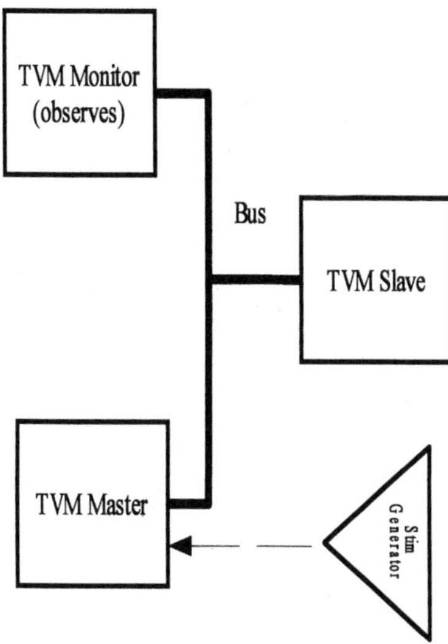

Figure 4-5. TVM architecture

The TVM master and the TVM slave are active participants on the bus where the master is used on the transmit side and the slave on the receive side. The monitor (bus monitor) is a passive observer and checks for the correctness of the bus protocol and could also record the transactions. The bus monitor has the following features:

- Check the protocol and log any errors into the log file. General error logging mechanism should be followed, indicating the type of error and the simulation time.

- Maintain a summary of the transactions. In general it should be able to report the total number of packets/commands, total number of errors etc.

- Have the ability to pass info to either continue with the simulation or stop simulation on error.

- Have the ability to record all the activity into the SST2 database

4.4.7 Packet/Cell Generators

The packet and cell generators are the primary stimulus to the chip. They should be implemented in a bus independent manner. For example, it should be possible to send any of the following packets/cells on either or all of the chip level or block level interfaces.

4.4.7.1 Packet Generation

For packet/cell generation for all application types for DUV the generation features will be designed to emulate "real life" network traffic. The generator should be able to control:

- Rate at which traffic is generated.
- Maximum rate.
- Minimum rate.
- Average rate.
- Sampling period over which the rate is determined.
- Burst traffic.
- Error injection.
- Start stop traffic at any time.
- Framer model for both input and output side of the chip.

4.4.8 CPU Command Generators

Either the CPU or Estimator can act as the master thereby initiating data transfer on the interface bus. Test sequences that exercise various scenarios need to be built. Commands are valid either during initialization time when no traffic has started or during traffic.

4.4.9 Packet/Cell Checkers

The packet/cell checkers check the stream at the lowest layer, which is the bit stream interface. Checkers need to be available for all the different types of packet/cell traffic. It should have the following features:

- Check bit level/field level errors on the packet.
- Check CRC.
- Port address mismatches.
- Check for right data transformation.
- Check for corruption on packets.

- Check for range of sizes and other parameters supported by the chip.

4.4.10 System Level Checker

At system level data path and control path checkers will be integrated together to predict the valid response for various test stimulus. Instead of just checking at input and output of chip, checking will be also be done at selected block interfaces which are of interest.

4.4.10.1 Data path checker

Data path checker for DUV system level will check for data transformation through the system. Data transformation for all the application types as they move from various blocks will be verified. Checker will check at following interfaces:
1)	From input interface to Lut output (SPII).
2)	Lut output to Segmentation output (OAM, Segmentation).
3)	Control path output to RAS output. (RAS).
4)	RAS output to SPI Out block. (SPI Output).

4.4.10.2 Control Path checker

Control Path checker will check for the correctness at PFQ enqueue and PFQ dequeue interface. This will be integrated at system level with all other four checkers to check the whole system level functionality. The approach for developing Control path checker will be bottom up. There will be at least an empty block checker function available for every block at system level verification environment. As the features evolve these will be filled in with the functional or cycle accurate block models.
More detailed methodology and flows for checker implementation are discussed in "methodology chapter".

4.4.11 Rate Monitors

The rate monitors perform traffic analysis and monitoring. They have the following features:

- Statistics collections on the received stream including packet/cell count, max rate, average rate, minimum rate etc.

- Monitor QOS as well as rate violations and log error messages and stop simulation if instructed.
- Log data streams into the SST2 database.
- Measure cell latency and jitter
- Feedback to chip on the SPI Bus based on rate violations and congestion
- Measure CBR, ABR and VBR for ATM cell traffic.

4.4.12 Simulation Infrastructure

4.4.12.1 Regression Scripts

All run scripts will be developed such that they could be run as regression tests on all available CPUs using LSF. Run scripts which automatically run the test and report pass/fail will be developed. The scripts will have the following features:

- Automatically run the tests with an option to specify the test list or all the tests.
- Option to turn the dump to enable waveform viewing. This is to speed up the test execution.
- Option to turn off any internal and external bus monitors or checkers.
- Report via mail/web the results of the regression tests. It should be able to inform the owner of the test when a test has passed/failed.
- Ability to turn on code/function coverage statistics database.
- Automatically update the TestStatus.xls document that has list of all the tests to be run for the DUV.
- Options supported in run script for various versions of SCV.
- Directives/options supported in run script for gate level simulation.

Refer to Regression chapter for other options.

4.4.12.2 Code and Function coverage

Code coverage will be run from top level once all the sanity tests are passing to estimate the additional code coverage added by system simulation assuming the block testing has already met the code coverage metrics requirement. This will slow down the simulation a lot, so set of tests from regression have to be selected that would be considered to cover most of different RTL code.

For functional coverage transaction explorer will be used.

4.4.12.3 Verification Metrics

Three metrics will be collected throughout the verification process. These are code coverage, function coverage and bug rate.

- Code Coverage/Function Coverage

  The coverage of the DUV RTL achieved by the verification will be monitored. From block level RTL will be accepted after 100% coverage.

  Functional coverage will be run during last phase of verification to find out how much chip functionality has been covered as defined in test status/description document and how much more tests to be added.

- Bug Rate

  The number of reported bugs still open will be reported on a weekly basis. An asymptotic decrease in this number after stable RTL has been achieved will indicate convergence on a clean design. This will define the exit criteria for the RTL.

4.4.12.4 Memory Modeler Integration

The general philosophy in this section is to catch the design errors as early as possible when they appear in the internal/external memory as opposed to appearing on the output possibly many thousand cycles later. The Denali memory modeler provides many hooks via the C api and need to be integrated into the SCV environment. Among the many checks which are envisioned to be useful are:

- Linked List Integrity Checking

  Linked lists are used in a number of blocks in the device. Amongst the many checks are linked list integrity checking as well as floating list elements.

- Assertion Checking

  Several system assertions are predefined (such as un initialized memory read, or multiple writes before read). In addition user defined assertions can also be introduced. The user defined assertions need to further specified.

- System memory setup where the system memory spans multiple devices.

- Memory Data Structures

  Memory data structures for various packets and cell classes are envisioned to help debugging.

- Node movement checks (should be very useful for PFQ/memory manager)

- These tests check for data integrity as the data moves from one list to another, such as active lists to the free lists. These are extensively used in blocks like PFQ.

4.4.12.5 Tools Used

Following table lists the tool usage:

Function	Tool Name
Transaction Based Verification	TestBuilder
Advanced Analysis (Lint Checking , Assertion Basedand Code coverage)	Hardware Lint Checking (HAL), 0-In, SureCov
Simulation	NCSim
Functional Coverage	Transaction Explorer
Compiler	G++ (2.9.5)
Class Library	STL
Revision Control	CVS
Bug Tracking	Problem Tracker
Regression	LSF
Memory Models	Denali

4.4.13 Verification Infrastructure

4.4.13.1 Driver TVMs

Driver TVMs are required to generate traffic at each of the system level or chip level interfaces. There will be TVM written that mimics the signal protocol for the interface to drive cells/packets.

4.4.13.2 Top Level Interfaces

Following top-level interface TVMs will be developed to drive traffic/stimulus into the chip. This will be driven on both Receive and transmit side.

- SPI4

- CSIX

- PL3
- PL2
- NPSI
- UTOPIA 2
- UTOPIA 3
- Framer

4.4.13.3 Chip configuration interface

CPU interface: This TVM will generate the CPU commands to setup the chip.

4.4.13.4 Block level interfaces

For those blocks where block level environment is required with checkers, driver interfaces will be needed.

4.4.13.5 Monitors

Monitors will be provided for all the interfaces for top level and for all blocks in the chip. This is to track the cell/packet progression in the chip. Besides tracking cell/packets, monitors can do some basic error checking for signals or protocols. The monitors also pass on the data to the checker.

4.4.13.6 Checkers

The goal is to automatically check all the features of DUT at system level simulation. Automatic checkers will be developed for most of the feature set of DUT. The cell transformation will be tracked and checked at all of these interfaces. Besides, cell transformation control features will be added in the specific checkers. Note that it is not guaranteed that having a checker for particular block at top level will result in verification environment and checking for block level verification. The partitioning of checker at top level is done to accommodate for various features of DUT that cannot be tested without referring to handshaking protocols at these interfaces. If only data transformation is to be tested than only one top level transformation checker is sufficient. The checker methodology is to achieve more automatic checking for the DUT. Ideally if there are more resources and time, checkers and block environments can be provided for each block. Following are the interfaces that will have checker.

- SPII
- SPIO
- LUT
- Segmentation
- Control path in and out
- Reassembly

4.4.13.7 Common infrastructure

This will have common code and data structures for generation, checker, monitors etc. All the run scripts, regression scripts, and post analysis scripts go here.

4.4.13.8 Generator

Common generator code. Methodology chapter discusses more details of the generator.

4.4.13.9 Checker

Common checker code. Methodology chapter discusses more details for the checker.

4.4.13.10 Central Messaging

Common message reporting code. Methodology chapter discusses more details for Central messaging.

4.4.13.11 Common Monitor Task

Common monitor task. This task has all the common code for the monitors at various chip interfaces for example common data format and error/protocol checks.

4.4.13.12 Chip Configuration Setup

Chip configuration setup will be on block-by-block basis. Each block setup will have an option to setup the offset from within block level address or top-level address. So, same block configuration can be used for both block level and top level. All block configurations will be in DUT/top/rtlModel/<block name> path.

4.4.13.13 Test Stimulus

Front end for writing tests will be provided. For DUT test stimulus API has to be written. Methodology chapter discusses more details for this.

4.4.14 System Testing

This section will cover the categories of testing to be done for the DUT and explain about how testing will be done. It covers the categories that cover all different set of features for DUT as defined on <date>. More categories and details will be added as more specs are available.

The detailed tests for each category will be in test description document written after all specs for the DUT are finalized and released.

4.4.15 Basic Sanity Testing

Chip Initialization

This is a quick set of tests to ensure that the chip has been initialized properly. The test is performed by reading the internal registers and memories via the CPU bus.

- Check the status register on every block to make sure that the correct block id is returned.
- Issue the block initialization commands for every block without waiting for the block to finish execution of the command.
- On finishing the command execution to each and every block, check that memory is initialized correctly. Initially set of ranges can be chosen for checking, for example only check lower, higher ranges and some middle ranges randomly.

4.4.16 Memory and Register Diagnostics

The goal of memory and register diagnostics is to make sure the right memory (both external ZBT memories and internal embedded memories) is instantiated. These tests could also be used for manufacturing tests at the test house.

The general algorithm is as follows:

For each block
    Check the internal registers of that block

For each external and embedded memory
  Get memory parameters (size, width, single port/dual port)
    Issue the CPU commands to initialize the chip to address the memory
      Apply the March-C algorithm
        Issue the appropriate pass/fail output for that memory to the log file

4.4.17 Application types sanity tests

For system level sanity tests, initially a set of tests will be defined for sending in traffic for each of the application types. This test list will be run first with every release of RTL to verify any major connectivity or overall functionality bug introduced because of block level changes. Various other features in the chip will be randomly selected for each of these types. This sanity list of tests that is run first with RTL release will keep growing as RTL becomes stabilized for adding more features randomly in initial sanity tests.

4.4.18 Data path Verification

Data path verification which is intended to look for legal data transformation. These tests include a single instance of the device.

It basically looks for the data transformation specified in the table below. All traffic is processed in order (no out of order processing needs to be performed) by data path. If necessary the control path can be activated, for control path-checking reordering of traffic will be checked too. The tests will be done for individual types and mix of following application types.

Data Transformation Tests

Traffic Type	Input	Output	Memory
ATM to ATM	ATM	ATM	ATM Cell
Packet to Packet	Packet	Packet	64B chunks
AAL5 Reassembly	ATM	Packet	ATM Cell
AAL5 Segmentation	Packet	ATM	ATM Cell
Fabric Reassembly	Switch Cell	Packet	64B chunks
Fabric Segmentation	Packet	Switch Cell	Switch cell
LFI AAL5 Reassembly	ATM	Packet	ATM Cell
LFI AAL5 Segmentation	Packet	ATM	ATM Cell
MLPPP Packet Reassembly(Def-fragmentaton)	Packet	Packet	64B Chunks
MLPPP Packet Segmentation(Fragmentation)	Packet	Packet	64B Chunks
AToM Mapping (Cell Packing)	ATM	Packet	ATM Cell
AToM De-mapping(cell un-packing)	Packet	ATM	ATM Cell
AToM AAL5 Reassembly	ATM	Packet	ATM Cell
AToM AAL5 Segmentation	Packet	ATM	ATM Cell
OAM Cells	ATM	ATM	ATM Cell

The data transformation tests check for the data transformation given in the above table. There will be option to turn off control path checking.

4.4.19 Chip Capacity Tests

These tests are designed to test whether the data path can support Maximum traffic on all the supported ports and also to check whether the merge is happening properly at high speeds.

•The test should generate 16k packets (maximum) on all supported ports with a million flows and check to see that no packets are dropped. It should also read the internal port counters (both input and output) in both segmentation and reassembly blocks to see that the packet count is the same as what the test expects.

•The test should generate packets which are one byte larger than the internal buffers used in memory manager, segmentation and reassembly. This will test corner cases where there are buffer overflows.

4.4.20 Framer Model

Framer model on input and output side of the chip. On output side checking will be done to see that there is no underflow and overflow of traffic. Model on input side will make sure that cell/packet bytes will be dropped according to real life scenario when framer sees back pressure from the device.(input side to check no bytes dropped unless over subscription of traffic or back pressure because of known chip configuration setup.)

4.4.21 Congestion Check

Fill all the FIFOs for congestion check.

4.4.22 Customer Applications

This section needs to be filled according to the feedback from the marketing team based on the expected customer setups for the device.

4.4.23 Free Buffer List checking

At the end of each test run free buffer list will be verified.

4.4.24 Overflow FIFOs

Test conditions will be identified for overflowing FIFOs for various blocks in DUT.

4.4.25 Backpressure

Backpressure will be asserted externally to the DUT and response verified. There will be also be internal back pressure conditions generated. The monitors will track any back pressure at the interfaces.

4.4.26 Negative Tests

For each category, set of negative tests will be written that will generate bad traffic along with the normal traffic. Checkers will expect any dropping/discarding of packets or bad marked cells/packets. Bad traffic will be mixed randomly – conditions like bad MTU, bad Length, bad CRC, bad SOP, EOP. Regression will be run once with normal traffic and than every test in test plan will run with bad traffic.

Certain bugs have more chance of getting detected earlier by running regression extensively with bad traffic.

4.4.27 Block Interface Testing

Block interfaces will be tested for specific cases. There will be tests dedicated to check certain handshaking protocol at block interfaces that seem to be difficult to create while running normal tests because of simulation time restrictions. For this reason separate environment for Control path top level will be created to be able to generate certain test conditions easily at control interface.

4.4.28 Chip feature testing

4.4.28.1 Buffer Management tests

BM block takes information about queue sizes for QID, TID, CID, port and class/color from control path and makes drop/mark/pass(color) decision. Control path(or SEG) drops if drop indication is asserted. Tests should generate discard conditions for all the elements for discard to occur in first cell of packet, mid packet or end of packet. All ranges of packets should be used for randomly asserting conditions for drop. By default no BM will be enable in any of the regression tests. Besides the set of tests just for BM, there will be option to enable the BM on various elements (QID, TID, CID, port and class/color) in the regression tests.

Tests will be setup to issue tear down connection command while BM is enabled.

- Stress tests in Buffer Management.
- Corner cases for Buffer Management.

4.4.28.2 Memory Controller

For various memory controller in the chip(data path, Memd, QMemd, Nmemd) , test conditions will be created to generate full at the memory interface. If it is difficult for generating full conditions with traffic complexity that simulation can support, then workarounds will be used to create such conditions at system level.

Accurate bandwidth calculation will be done for various test conditions at mem controller interfaces.

CPU response time at frequency 266Mhz, 300 Mhz and 200Mhz will be verified.

- Corner cases for Mem controller.

  While full is asserted for memory controller, send in negative traffic combinations.

- Stress cases for Mem controller.

4.4.28.3 Performance tests

Testing will be done to see following performance requirements for DUT are met:

- Full-duplex maximum rate supported.
- For ATM->PKT should support Incoming traffic rate maximum supported and outgoing traffic rate maximum supported.
- For PKT->ATM should support Incoming traffic rate twice maximum supported and outgoing traffic rate maximum supported.
- Check for max rate for two Byte(second byte size 1)packet.

4.4.28.4 Statistics

Tests will be written to collect input and output statistics per QID for DUT. Besides specific tests, every test in regression will collect and verify statistics at the end of test.

4.4.28.5 Multicast

Multicast feature will be tested for various combinations of roots and leafs mixing it with unicast QIDs. Besides mixing with unicast FIDs, negative traffic will also be send in for discard in multicast tests.

4.4.28.6 Policing

Tests will be written to check byte or cell accurate policing for ATM or packets using single or dual bucket policing algorithms.

4.4.28.7 Shaper

Specific tests will be written for checking shaper functionality at top level. Also, there will be option in each test running in regression to randomize shaping parameters and automatically check it.

4.4.28.8 Input and output interfaces

Tests will be set to randomly send traffic on all supported input interfaces to DUT – Utopia 2, Utopia3, PL2, PL3, SPI4, and Framer interface.

Pass through mode for interfaces will be enabled for some tests where output interface provides backpressure to the input interface.

Chip will be setup in re-circulation mode of operation and response will be verified for backpressure asserted by re-circulation block on input interface.

4.4.28.9 CPU port

Some set of tests sending traffic on CPU port and injecting traffic into CPU mode. Mix of traffic with normal traffic.

4.4.28.10 Control path testing

Separate top-level control path verification will be used to do control path tests conditions that are more difficult to generate from system level because of simulation restrictions with using chip to its full capacity.

4.4.28.11 LFI (Link fragmentation and interleaving)

Test for LFI features mainly data transformation and timeout and sequence number checking. Verification infrastructure will be written for generating various LFI format traffic.

4.4.28.12 ATOM (Any Transport over MPLS)

Test for ATOM features mainly data transformation and timeout checking. Verification infrastructure will be developed for generating various ATOM format traffic.

4.4.29 Random Verification

- All the generation will be random from the stimulus file except for the fields/ parameters specified to be directed.
- Tests will be initially written for directed testing and will have option for random generation based on the seed
- All randomization will be within constraints.

- All don't care signals/mems will be driven randomly for each test run.
- In this case the rate generation could be a random quantity bounded by peak and base rates across all the ports.
- Mixture of differing protocol generators on the different input ports, randomly generated.

4.4.30 Corner Case Verification

This section will cover corner case tests identified for simulation.

4.4.31 Protocol Compliance checking

All the protocol compliance checking to be done will be added to this section.

4.4.32 Testing not done in simulation

This section will cover for each block what is the testing that is not done.

The full range up to upper limits for all supported ranges in DUT for different features will not be tested in simulation. It's not possible in simulation to be able to setup and check full ranges because of compute memory and simulation tool limitations. All the stress testing for chip will be done in FPGA. Proposal will be made to convert verification tests stimulus in the form that tests can drive the FPGA platform so that in later phase verification team can do some of stress testing on the FPGA board.

4.4.33 Test Execution Strategy

In the sub- sections below is covered the test execution strategy based on the test schedule. Each of the section talks about what should be covered. It is like a template and not an actual example.

4.4.33.1 Testing Schedule

Prepare a testing schedule to reflect various types of tests and duration of each test. This should reflect the personnel involved in the test effort. In testing schedule include the Documentation review, test scripts, data preparation, test execution, output review/analysis, and sign off criteria for the tests.

4.4.33.2 Equipment requirement

Provide a chart or listing of period of usage and quantity required for each equipment employed throughout the verification process. While regression is running - you might need 100s of licenses and cpu slots for concurrent test running over a short period of time

4.4.33.3 Tool requirement

Identify any tools required in support of testing and for execution of tests.

4.4.33.4 Testing Progress

This section will describe how progress would be made from one phase of testing to another. You can give the list of tests to be run for Sanity checking for one-two hours when RTL is released. Next is list of directed, random and constrained random test categories and lists and the final metrics that guides the sign off.

4.4.33.5 Procedure to run the test

Describe various procedure to run the test step by step procedure. Setup scripts, environment, machine, lsf setup, initialization. What switches in run script decide how chip is initialized with each test, termination of testate is test stopped and logged based on what criteria.

4.4.33.6 Result Analysis

How to analyze output results and notify them? Who will send the reports to whom and how the results will be analyzed by different level of people for example designers can look at log files, higher level management look at charts etc.

Overall Identifying which tests contribute to overall testing strategy and which tests duplicate other tests can save considerable time and effort during regression runs. tests that are subtests of others.

4.5   Tests Status/Description document

Test description document is required for detailed listing of tests, to get the coverage metrics from each test category and track daily progress during test execution phase.

Once RTL is released, tests listed in this document run in the regression on daily basis until all the tests are passed and functional coverage for each test has reached to the agreeable stage. Each test is a feature to be tested and it is implemented in a constrained randomized manner so that every time it runs in a regression it picks up the parameters for the feature to be tested in random manner. While this test runs in regression the seed gets recorded for the random parameters in the log file so that test can be rerun in case of failure. Functional coverage can be used to find out the percentage of the ranges covered.

Detailed testplan can be written in XL spreadsheet format. Utilities scripts can be written to automatically update status in the spreadsheet based on the regression runs.

An example is given in figure 4-6 for the basic spread sheet to be maintained, other spreadsheets are detail test names with their status and summary/chart to be updated on daily basis.

- Features/category is the feature name to be covered from the functional spec.

- #subtests is number of tests that belong to that category where each test is specific feature to be tested. It can be combination of directed or random tests or constrained random tests.

- Priority is the priority of the test in relation with the others based on the application needs for the DUT.

- Owner is the name of the engineer who owns implementation and execution of the test.

- Status is the current status of test. It can be test not implemented, test not run yet, test failed, or test passed

- Coverage is how much coverage has been achieved for the particular category to be tested.

Feature	#subtests	priority	Owner	Status	Coverage
Protocol Compliance checks					
Chip capacity tests					
Interface testing					
Memory Interface					
Performance					
Control path					
Data path Segmentation/R eassembly					

Figure 4-6. Test description document

Test description document is living document until sign off on all the functionality covered according to the tests mentioned in the test plan.

4.6   Summary

This chapter gives an example of an effective test planning strategy. Test plan can be split into System level plan/methodology document and test description document. The methodology document can highlight the main features to be tested and methods to test them. Test description document lists the details tests and maintains the coverage metrics until the sign off is met. The example has been provided for some of the cases from the network chip reference design.

CHAPTER 5

Testbench Concepts using SystemC

Testbench components are fairly well known these days. They typically include tests, test harnesses, monitors, master, slaves, etc. In this chapter let us define these different components and give examples of how to implement them efficiently using SystemC. Lets us also take a quick look at the key concepts of unified verification methodology, and see how it can be applied to and be part of the creation of an efficient Verification environment.

5.1　Introduction

Lets look at the different modeling methods used in an advanced verification environment

5.1.1　Modeling Methods

For transaction based verification, a method must be chosen to represent the system under test, injecting stimulus, and analyzing the results. All of these are dependent on each other, and they all require both tools and methodology to work efficiently.

The representation of the design may be a low-level form such as an RTL model, or in one of several more abstract forms, such as a transaction-level model (TLM) Similarly, the entire system may be represented in a single abstract form, or any mix of components. The more abstract models tend to be significantly faster to write and execute, but with less detail. It is not uncommon for projects to mix these models in various ways as the project progresses.

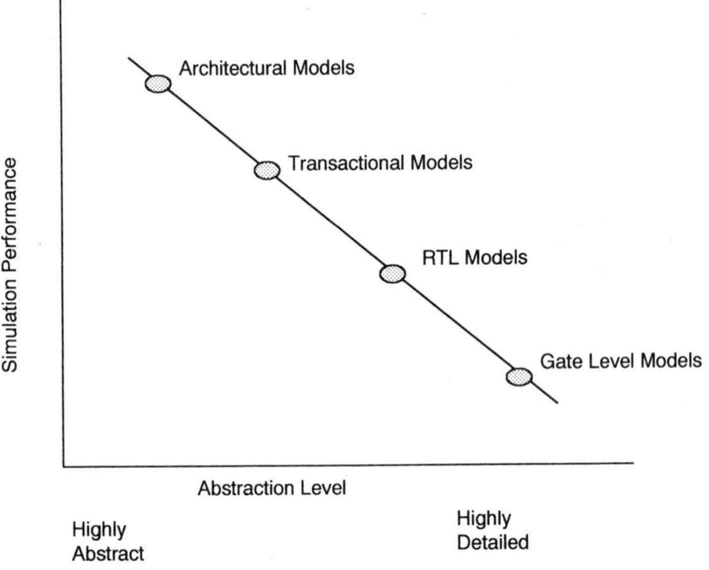

Figure 5-1. Model abstraction Model

5.1.2 Transaction Level Modeling

Transactions provide an abstract model of a signal-level bus interface that will be used to communicate between devices. Rather than providing all of the signal-level protocol details, the transaction-level interface focuses on the data transfer and handshaking between blocks. This higher abstraction level may be easier to implement early in the project, permitting system-level model to be run earlier in the project cycle with greater efficiency.

A transaction level model (TLM) provides a similar relationship to RTL models. Where an RTL model provides the details necessary to synthesize and build a device, a transaction-level model is much more abstract. A TLM models the architectural behavior of a block, accepting input transactions and driving output transactions in the same way that an RTL model would, but with much less internal structure. The TLM is focused on the behavior of the block without any of the bus-level protocol or implementation details. As a result, the TLM tends to have significantly less code, is easier to write, and faster to simulate.

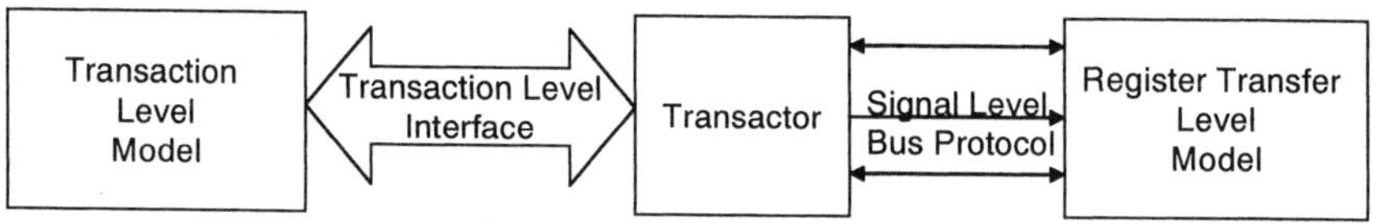

Figure 5-2. Transactions enable communication between different levels of abstraction

5.1.3 Block level Verification using TLM

As a particular module or block is being refined from an abstract to an RTL level, the TLM can be used to help verify that the RTL block performs the same functions as the original TLM. This can be done in both a stand-alone environment, as well as in a larger sub-system or system-level setting.

At the block level, a standalone verification environment can use the TLM as a reference model to determine correct behavior of the block. At this level, the RTL block is communicating with the verification at a signal level. Transactors are needed to monitor the signal-level busses and convert them to transactions for the TLM. Figure 5-3 shows a basic diagram of what this will look like.

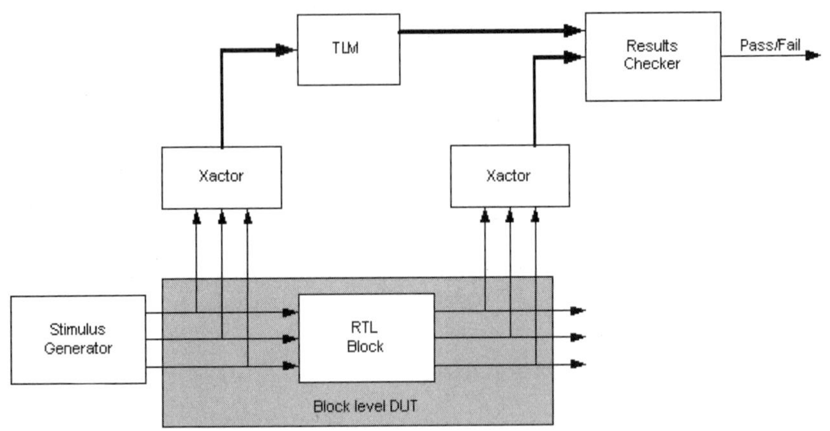

Figure 5-3. Block Level RTL Verification with a TLM

This approach to verifying an RTL block may save a significant amount of time, since a results checker now needs only to compare the outputs between the TLM and the RTL block. How that comparison is done will depend on the accuracy of the TLM. This may vary from an extremely accurate model, that will mimic the output of the RTL, to a very abstract model that will provide the same algorithmic modifications, but with differences in output times, and potentially differences in the order of the output transactions.

5.1.3.1 Building a Results Checker

The results checker must compare the transactions arriving from the RTL block with those from the TLM. Conceptually this is a fairly straightforward task, since predictions are not necessary. However, because the TLM is a more abstract model, and not an exact functional replica of the RTL, there may be differences between the models. The results checker must be able to determine if the transactions received from these two models are functionally equivalent.

Notice that the response checker could be built to compare signal-level interfaces or transaction-level interfaces. The advantage to comparing at the signal level is that the comparisons may be easier to do in the case where the signals match exactly. However, signal-level comparisons may be somewhat slower, since there is more detail involved. More importantly, any mismatch will probably be reported at the signal level rather than the transaction level. It may take more time to determine which operation mismatched, and what the cause of the mismatch was. As a result, transaction-level checkers are generally preferred. The examples in this document show transaction-level checking.

5.1.3.2 Cycle-Accurate Model

When the TLM is a cycle-accurate model, then each transaction from the TLM will exactly match those from the RTL. This makes the job of the results checker easier, since comparisons are direct. Figure 4 illustrates the comparisons needed for a cycle accurate model

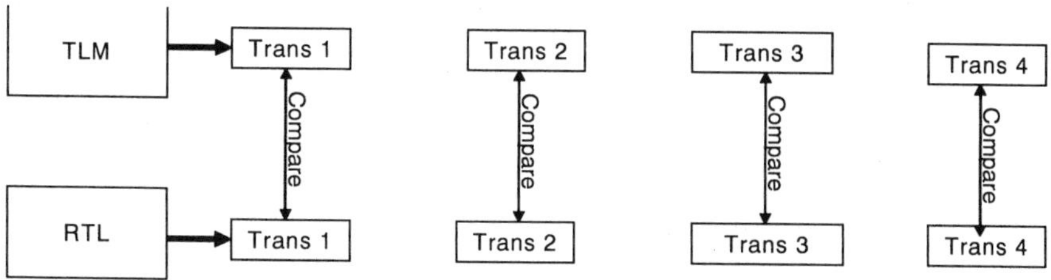

Figure 5-4. Cycle Accurate Results Checking

5.1.3.3 Order-Accurate Model

Often the TLM is significantly more abstract in nature than the RTL. In that case, it is not uncommon for the transaction-level model to issue transactions at different times than the RTL does. Each transaction will still be equivalent, but the response checker must be able to compare the results even when they don't match up in time. Figure 5-5 shows an example of transactions arriving in order, but out of time alignment. This type of behavior will happen if a TLM is not cycle accurate, or has no timing information at all. Since a TLM is a more abstract model, it does not have the same level of detailed information that is in the RTL model. A TLM can be written that estimates the delay of the RTL, but the actual delays may not even be known when the TLM is written. Similarly, a TLM that has no cycle timing information at all will issue transactions immediately, while the RTL model may require many cycles to issue the same transaction

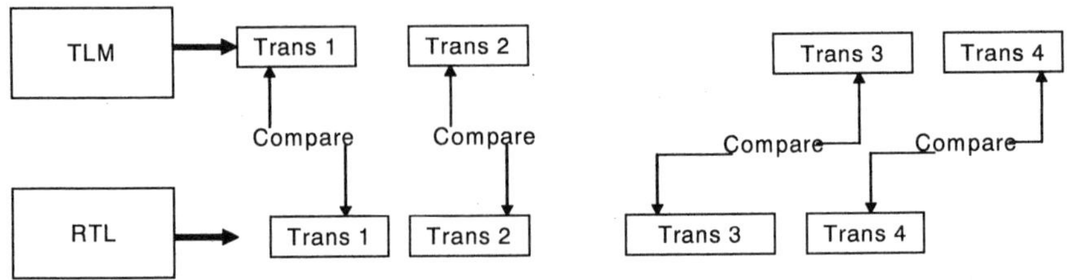

Figure 5-5. Order Accurate Results Checking

5.1.3.4 Transaction Accurate Model

In some systems, the ordering of transactions is modified in the RTL in ways that are not easy to predict within a transactional model. Out of order transactions are likely to occur in systems where caching or queuing is occurring within the model. An algorithm within the model is responsible for determining which entry to select. In some cases, the algorithm is based at least partially on a pseudo random number. Other times, the algorithm used to select entries is very specific. In either case, the transaction-level model may not have all of the necessary information, such as the pseudo random number generator, or the abstraction level is too high to exactly model the selection algorithm used in the RTL.

In these cases, the order of transactions may also be different between the two models as shown in Figure5- 6. At this point, a results checker will need to store transactions from one transactor until a matching transaction is received from the other. While this requires some more logic on the part of the results checker, it is still generally far simpler than attempting to predict behavior without the TLM. Even the most abstract TLM will simplify the task of results checking

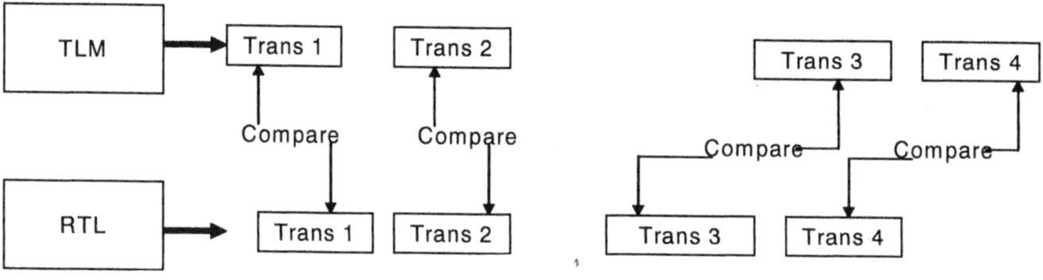

Figure 5-6. Transaction Accurate Results Checking

The transaction based results checker is an idealized model, where all interactions can be expressed as transactions. In some cases, this is a valid model, particularly for blocks that are only connected to well defined busses, such as an AMBA bus. In other cases, the RTL block may have more customized interfaces with individual control signals that are passed between blocks.   In cases where individual signals are passed between blocks, how the results checking is done may depend on the implementation of the transaction level model. If the TLM has been implemented to reproduce any signal-level communications, then the results checking is straight-forward. If the abstract model does not reproduce all signals, then some additional thought must go into the RTL-level testbench.

5.1.3.5 Complex Results Checking

As more complex transactions are considered, there are times that the abstract model may provide accuracy at a high level, but the outputs of the TLM may not have an exact relationship to those of the RTL. One example of this occurs in some communications devices, where a large data payload is being sent. The TLM may provide the complete payload, where the actual implementation may decide to break the data up into smaller fragments for efficiency or protocol reasons. This is one example where there is high-level agreement between the TLM and the RTL but the actual transactions do not agree. Figure 5-7 shows an example of frag-mented packets.

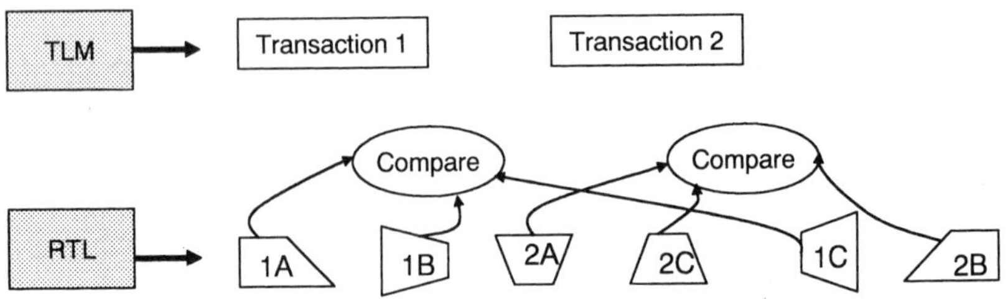

Figure 5-7. Fragmentation of Data

This type of situation may happen when there is a protocol layer that is higher than what the RTL provides, or the TLM is operating at an abstraction level that does not actually exist in the actual implementation. In either case, the results checking must now be translating the received transactions before they can be compared. In some cases, there may be a TLM, or a portion of a TLM that can provide that translation. In other cases, the data must be collected and compared using additional logic.

Most solutions require that the response checker be able to provide more than a basic comparison of output transactions from the TLM. Comparison logic is now required as part of the checking.

5.1.3.6 Order Requirements

In figures 5-6 and figure 5-7, the ordering of operations has been modified by the RTL. In some cases, how the ordering changed is not important as long as the correct data is sent. In other cases, the order of information is as important as the data being checked. High performance systems will often put a great deal of effort into determining the optimal ordering of data. In these types of systems, it is not sufficient to simply check that the data was received. It is also important to determine that the data was received in the correct order.

The ordering mechanisms built into the RTL are often very detailed, including pseudo-random number generators, and state-sensitive logic. These are the types of mechanisms that are not identical when they are abstracted. As a result, the TLMs will rarely mimic the ordering mechanisms of the more detailed RTL implementation.

Because of this, result checking cannot rely on the TLM to determine correct operation. There are generally several other options available to the results checker to determine if the ordering was correct. How this is done will depend almost entirely on the specific system, but a results checker may want to monitor certain aspects of the RTL, or the transactions entering the RTL to determine if the reordering was correct.

The results checking could verify that the order of every transaction was correct. In this case, it may need to monitor some internal states of the RTL, such as the number of transactions in queues. This is a highly detailed approach for a results checker. The advantage is that any aberration would be caught quickly. The disadvantage is that it takes time to build such a results checker, and there is a risk of re-implementing a design error in the checker.

A second approach that may work for some systems is to take a more statistical approach to order checking. If the goal of the results checking is understood, then the checker can measure the critical performance issues of each transaction and determine if the ordering is working correctly based on averages of many transactions. For example, if a device is built with a quality of service capability, then some high-priority transactions will be expected to pass through the system faster than low priority transactions. The results checker can measure the latency of the high and low priority transactions and determine if the ordering is correct based on the average latency of each type of transaction. The advantages of this method is that it is more abstract, easier to implement, and more accurately checks for the overall goal of the system. Among the disadvantages is that there is no clear error report to indicate that a specific transaction was not ordered correctly. All the results checking can do is report that an error threshold has been exceeded.

5.2   Unified Verification Methodology(UVM).

A unified verification environment allows models at various abstraction levels to be combined in a single system level model. Transactors provide the translation between a transaction level communication and the signal level protocol to allow TLMs and RTL models to communicate with each other. With the use of transactors, TLMs and RTL can be mixed in any type of environment.

The methodology is centered on the creation and use of a transaction-level golden representation of the design and verification environment called the Functional Virtual Prototype. The methodology encompasses all phases of the verification process and crosses all design domains. Utilizing the Unified Verification Methodology will enable development teams to attain their verification goals on time.

While focusing on the verification of System-on-Chips (SoCs) the methodology also encompasses verification of individual subsystems. Large application-specific digital or analog designs are often first developed as standalone components and later used as SoC subsystems. The Unified Verification Methodology can be applied in whole to a SoC design or in parts for more application-specific designs. Different designs and design teams will emphasize different aspects of a methodology. The Unified Verification Methodology will produce the greatest gains in speed and efficiency when used in a complete top-down integrated manner. It is understood that a complete top-down flow may not always be feasible for a number of different reasons. Thus the methodology is flexible in providing for both top- down and bottom-up approaches to developing subsystems while still providing an efficient top-down verification methodology.

While it may be impractical for verification teams to move directly to a top-down unified methodology from their existing methodology, this methodology provides a standard for verification teams to work towards. Some key UVM concepts are discussed in following sections.

5.2.1 Functional Virtual Prototype

A Functional Virtual Prototype (FVP) is a golden functional representation of the complete SoC and its testbench.

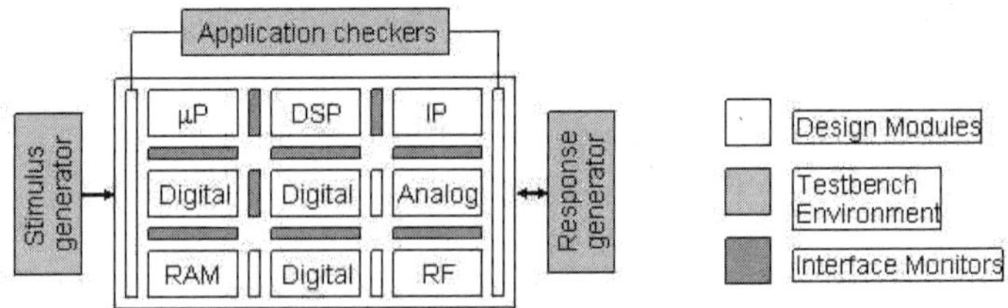

Figure 5-8. Functional Virtual Prototype Example

The FVP exists in many different levels of abstraction throughout the design. The transaction-level FVP is created early in the design process by the SoC verification team working closely with the architects. The FVP is termed transaction-level because the communications backplane that connects the functional models transfers information in the form of transactions. As development progresses implementation-level models provided by the subsystem teams replace transaction-level models creating a FVP with mixed levels of abstraction. During this process the communications backplane of the FVP changes to transferring information in the form of signals. Once all of the transaction-level models are replaced with implementation blocks the FVP is considered an implementation-level FVP.

The FVP unifies the verification of different design domains and processes by providing a source of common reference models as show in figure 5-9 below.   The design in the figure 5-9 begins with four blocks represented first as a transaction-level FVP. Each block is developed standalone using the TLM from the FVP in the testbench. The implementation-level blocks then replace the TLMs in the FVP creating the final implementation-level FVP.

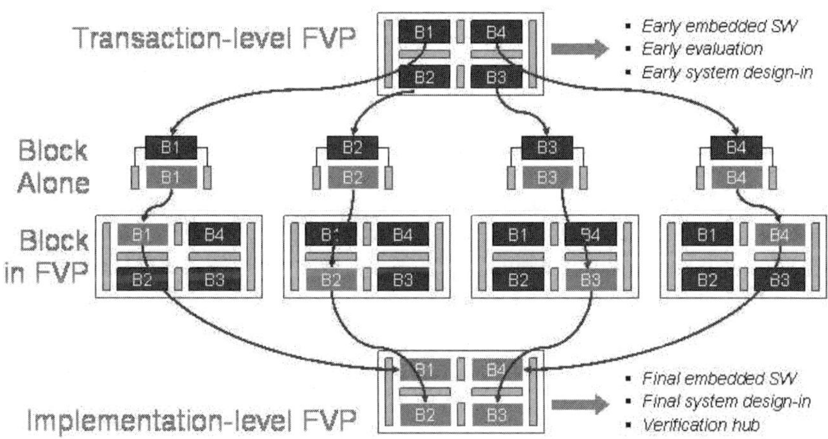

Figure 5-9.

The transaction-level FVP consists of Transaction-Level Models (TLM), transaction-level testbench components and mixed-signal instrumentation. TLMs model each of the functional blocks within the design. The initial degree of fidelity of the TLM to the implementation is dependent on the needs of the software and system development teams. If the TLM is to be reused for subsystem verification then the implementation fidelity is also dependent on the verification team's needs. In most cases the TLMs are behavioral code wrapped in a transaction-level shell for interfacing to other blocks as shown below in figure 5-10.

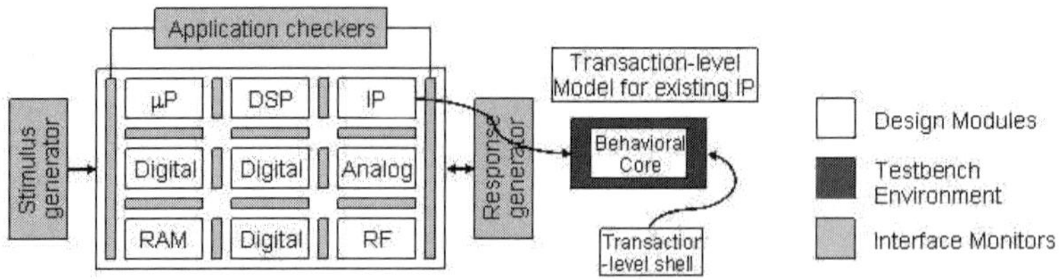

Figure 5-10.

Each subsystem has its own TLM defined at the SoC partition. The subsystem team may further partition the subsystem TLM into implementation-level block TLMs to be used for block-level verification components.

The FVP also contains testbench components including stimulus generators, response checkers, interface monitors and mixed-signal instrumentation. The FVP includes these components to provide a comparison environment for subsystems to develop against and to allow for testbench reuse. The testbench components are all at the transaction-level providing high performance. Mixed-signal Instrumentation includes specialized stimulus and response analysis monitors used to interpret mixed signaling. A test suite is included with the FVP. This test suite is run on the FVP as each subsystem implementation is integrated into the SoC. The subsystem teams use this environment to test individual blocks as they are developed identifying integration issues with other subsystems early in the process.

The process of integrating implementation-level blocks into a transaction-level FVP is facilitated by the use of transactors. Transactors translate information between the transaction-level and the signal-level. An example of using transactors to integrate a single block into the FVP is show below in figure 5-11. In this example the implementation-level block is surrounded by transactors along with interface monitors at the signal-level. This entire group of elements replaces the Digital TLM in the FVP allowing for a mixed-level simulation.

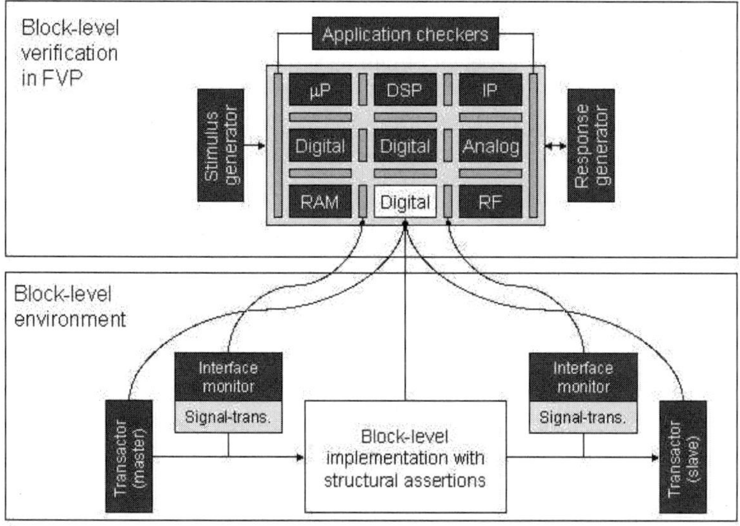

Figure 5-11.

The FVP serves several critical roles in the methodology:

- An unambiguous executable specification
- A fast executable model for early embedded software development
- A model for performing architectural trade-off
- An early handoff vehicle to system development teams
- The reference for defining transaction coverage requirements
- The source for subsystem-level golden reference models
- A golden top-level verification environment and integration vehicle

5.2.2   Transactions

Transactions improve the efficiency and performance of the Unified Verification Methodology by raising the level of abstraction from the signal level. A transaction is the basic representation for the exchange of information between two blocks. Put simply, a transaction is the representation of a collection of signals and signal transitions between two blocks in a form that is easily understood by the test writer and debugger. A transaction can be as simple as a single data write operation or a sequences of transactions can be linked together to represent a complex transaction such as transfer of an IP packet. The following figure 5-12 shows a simple basic transaction data transfer (represented as B) linked together to form a Frame, further linked together to form a Packet and finally a Session.

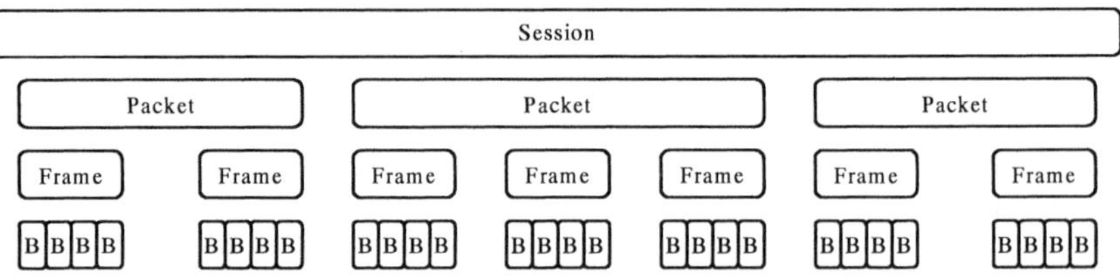

Figure 5-12.

Transactions are used throughout the Unified Verification Methodology. A Transaction Taxonomy is created early in the development process to assure the common use of transaction types throughout the process. This provides for reuse of test-bench elements and analysis tools with a common transaction reference. The Transaction Taxonomy specifies the layers of transactions from the simplest building block to the most complex protocol. The Transaction Taxonomy is created by defining the appropriate levels the test writers and debuggers will be most efficient operating at. All models and tools are written to these levels of abstraction depending on their use. The table 5-1 gives the transaction taxonomy for the previous diagram's transactions.

Table 5-1.

Level	Data Unit	Operations	Fields
Interface	Byte	Send, Receive, Gap	Bits
Unit	Frame	Assemble, Segment, Address, Switch	Preamble, Data, FCS
Feature	Packet	Encapsule, Retry, Ack, Route	Header, Address, Data
Application	Session	Initiate, Transmit, Complete	Stream

The use of transactions in the Unified Verification Methodology improves the speed and efficiency of the verification process in several ways:

- Provides increased simulation performance for the transaction-based FVP.
- Allows the test writer to create tests in a more efficient manner by removing the details of low-level signaling.
- Simplifies the debug process by presenting the engineer information in a manner that is easy to interpret.
- Provides increased simulation performance for hardware-based acceleration.
- Allows easy collection and analysis of interface-level coverage information. Information in Table 5-1 can easily form the basis for creating cover-groups and coverage model design discussed in detail in Functional Coverage chapter 8.

5.2.3 Assertions

Assertions are used to capture the intent of the architect or implementer as they develop the system. Verification tools can be used to verify the assertions either in a dynamic manner using simulation or in a static manner with formal mathematical techniques. Assertions are created in the Unified Verification Methodology whenever design or architecture information is captured. These assertions are then used throughout the verification process to efficiently verify the design. There are three types of assertions used in the Unified Verification Methodology. First, Application Assertions are used to prove architectural properties such as fairness and deadlocks. Second, Interface Assertions are used to check the protocol of interfaces between blocks. Finally, Structural Assertions are used to verify low-level internal structures within an implementation such as FIFO overflow or incorrect FSM transitions.

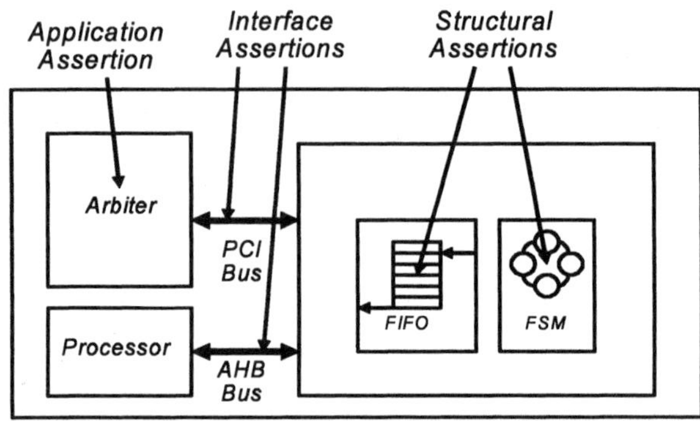

Figure 5-13.

The use of assertions in the Unified Verification Methodology improves the speed and efficiency of the verification process in several ways:

- Speed the time to locating difficult bugs by identifying where in a design the bug first appears.

- Automate the instrumentation of monitors and checkers within the design.

- Quickly identify when stimulus or interfacing blocks are not behaving as the implementer intended.

- Identify bugs that did not propagate to checkers or monitors.

- Protocol violations can be detected even if they do not cause a functional error.

- Provide feedback to stimulus generators to modify their operation as the test progresses.

5.2.4 Considerations in creating an FVP

The first step in creating a FVP is determining the intended use of the FVP. Each design and each development process is unique and has its own set of challenges. The FVP is not intended to be a "one size fits all" solution. While the accuracy of the FVP must always be correct the degree to which the models in the FVP represent the implementation of the design, otherwise known as the fidelity of the design, may vary. The detail and fidelity of the FVP is determined by its intended use. There are three main uses of a FVP:

1. Executable model for software development

2. Reference model for subsystem development and integration

3. Model for the team that will use the developed design in the final system, often known as the design-in team.

Each of these three uses has different requirements for the FVP and how it will be developed. Development team must determine which of these uses are relevant to their design and what priority should be placed on each. The following sections describe the considerations in creating a FVP for each intended use.

5.2.4.1 FVP use in software development

The ability to develop and debug software on an executable model can be a large time savings for a project. The amount of time saved is dependant on the resources available along with the type and amount of software needed to be developed. Many software teams today are resource limited and are not available to begin work on a new project until the hardware has been developed. In these cases the benefit of the FVP as an executable model are limited. If resources are available then the type of software that is needed to be developed plays a major role in the development of the FVP and the benefits it provides. Three basic types of software systems are discussed below:

1. Service processor applications: Designs where software controls a service processor to handle basic start-up and maintenance functions requires the least amount of fidelity in the FVP. In these designs the software needs read and write access to registers and memory space within the design to perform operations such as configuration of mode registers, error handling and collection of runtime statistics. To support this software development the FVP needs to closely model the software accessible registers within the system and provide basic register and memory functions. The algorithms, data-paths and processes within the blocks can be very high level and implementation non-specific.

2. User interface applications: Designs where software controls the processor to allow the user to control and monitor the operation of the system requires greater fidelity of the FVP than a service processor application. In these designs the software provides a user interface to allow the user to change operation of the system and to provide important run-time information on the status of the system. To support this software development the FVP needs to closely model the software accessible registers within the system and be able to monitor run-time events as they occur within the system. The algorithms, data-paths and processes within the blocks need to provide enough visibility and control as required by the software UI.

3. Real-time software applications: Designs where software is directly involved in the functional processes and algorithms of the system require the most amount of fidelity in the FVP. In these designs the software it tightly coupled with the hardware to provide the functionality of the system. To support this software development the FVP needs to closely model the software accessible registers and memory along with closely modeling the algorithms, data-paths and processes within the blocks. Often this modeling has already been done in the architecture stage of the design.

Understanding the amount of software to be developed as well as the application should provide the FVP developer with a good understanding the required fidelity of the FVP for its use in software development.

5.2.4.2 FVP use in subsystem development and integration

When creating the FVP for a system it is important to understand the specific needs of individual block development before beginning. Individual blocks within the FVP either already exist in an implementation form, such as third party IP or reused cores, or they entail new development. Existing designs may not already have a TLM built for it. In these cases the FVP team must create a TLM for the existing implementation that is used only for integration purposes. The TLMs of these existing designs should concentrate on fidelity at the interface level and abstractly model the datapath and control processes of the design.

FVP blocks that are to be developed may require more detailed modeling. One important factor to consider is if the block will use the TLM in a top down manner to verify the individual sub-blocks as they are designed. If this is the case then care should be taken in creating a TLM that correctly partitions the functions as they will be partitioned in the design. If the development team plans to only use the TLM from the FVP to do full sub-system verification then internal partitioning is not necessary. FVP modeling of mixed signal and Analog blocks is described in section 5.2.5.4.

5.2.4.3 FVP use in system design-In

Most systems today are designed by a group of design chain partners. The design that the FVP is created for will most likely be designed into a larger system by a different team or customer. The FVP provides these design-in teams a model to begin developing their systems before the implementation is ready. In some cases these design chain partners will use the FVP as part of their larger system model to test functionality and performance. The FVP also provides an excellent vehicle for demonstrating to customers and third-parties that design is progressing as expected. FVP developers need to understand the needs of their design chain partners for the FVP. The FVP may require enough fidelity for the design-in team to develop their models as well as system software.

5.2.5   Creating the FVP

5.2.5.1   Modeling Language

The FVP is first defined as a transaction-level in the system-design stage and then is refined and used throughout the development process. It is important that the FVP be written in a language that supports system modeling and also can be used throughout the development cycle. Given these requirements SystemC qualifies as one of the best language to use to develop the FVP. This section is not intended to discuss how to create a TLM with SystemC. This section will discuss the basic steps in creating a model assuming SystemC is the language used.

5.2.5.2   Creating a FVP TLM

The FVP is created in a top down fashion following the partitioning of the system architecture. The first step is to identify the modules for the system and the correct hierarchical partitioning. Next a hollow shell of each module should be created with a name that will match the implementation modules name. External ports for each of the modules should then be defined and added to the individual modules. Consistent naming between the model and implementation is important to facilitate future integration.

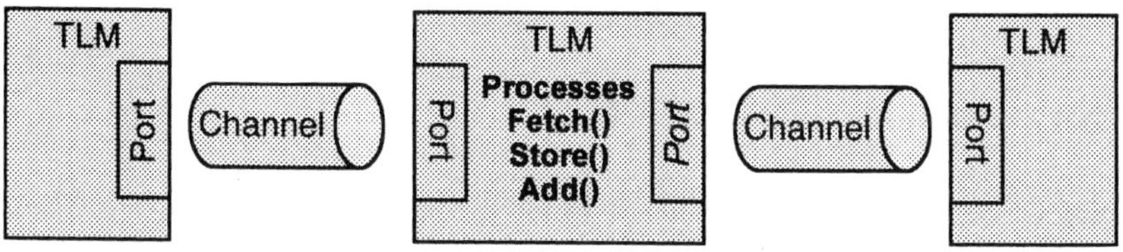

Figure 5-14.

After the modules have been defined the developer should define the channels that the modules will be interfaced to. Again careful consideration and planning is necessary to provide common interfaces that can be used throughout the process. Once the modules and channels are defined each of the modules has their individual processes defined. The modules are defined to be functionally correct representations of their corresponding implementation. Separate threads for parallel processes are used with the state stored in member variables.

5.2.5.3 Creating a TLM from a behavioral model

Often a subsystem is already implemented when the FVP is being created. In these cases a TLM may not have been created for the subsystem when it was developed or the only model available is a behavioral model. A TLM can be created from a behavioral model by surrounding the behavioral or implementation model with a transaction-level wrapper. This wrapper translates the behavioral interface into a transaction-level which can be used in the FVP. The transaction-level wrapper captures the output of the model and calls the SystemC transaction functions across the channels. The transaction level wrapper also receives transactions across channels from other subsystems and converts the transaction information to a form the behavioral model can utilize. This is shown in the figure 5-15.

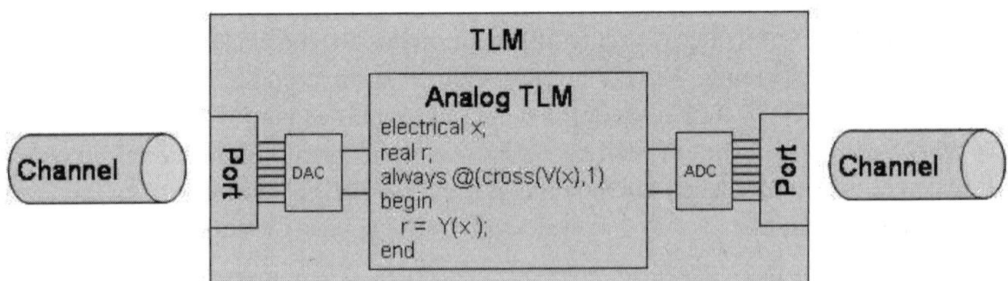

Figure 5-15.

5.2.5.4 Creating a Analog FVP TLM

Analog designs do not operate at the traditionally defined transaction-level. Analog designs operate in a continuous-time manner versus a discrete-time operation. A transaction in the continuous time domain can be thought of as a sequenced response over the time unit. Thus, a TLM model of an analog component looks very similar to a standard continuous time model. Analog algorithms are often first represented as high-level C functions and are refined down to behavioral models that represent the implementation of the design. These models accurately reflect the continuous-time nature of an analog circuit.

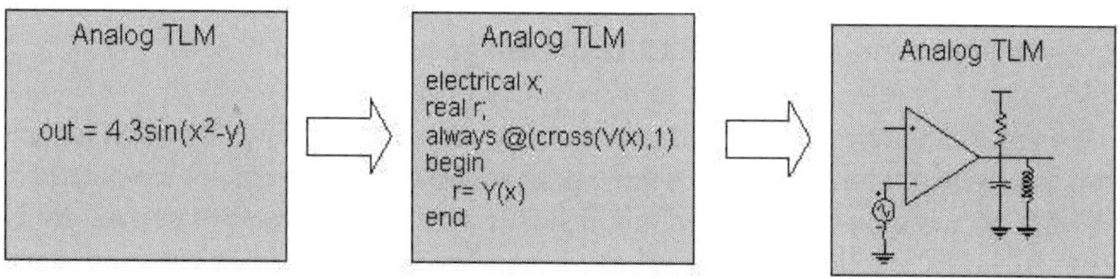

Figure 5-16.

The important factor in creating analog models for an FVP is accurately modeling how they will interface to the other subsystem models. The interfacing between analog blocks is done in the continuous domain. The interfacing of digital and analog subsystems is usually through a basic converter (Digital to Analog or Analog to Digital). This D-to-A or A-to-D converter transfers the continuous-time algorithm to a signal-level digital interface. This digital interface can then be translated to a transaction interface to allow for connection to digital TLMs.

5.2.5.5 Creating an Algorithmic-Digital TLM

Algorithmic digital subsystems are similar to analog subsystems in that they are first modeled as continuous algorithms. Algorithmic digital subsystems are refined down first to fixed-point models and then to discrete-digital models. These subsystems usually interface between several different domains. Often they interface to the analog subsystem where at first the interface is modeled in a continuous-time nature. Then once a D-to-A or A-to-D converter is placed at the interface the algorithmic subsystem is modeled as a fixed-point representation.

Figure 5-17.

Algorithmic digital subsystems also often interface to the control-digital subsystems in the FVP. This interface is how the subsystem gets configured for operation and provides the user with statistical information. This interface is usually a standard bus interface and is easily modeled as a TLM.

5.2.5.6 Creating stimulus and response generators

Once the FVP TLMs are created the model needs to be verified. Stimulus generators are created to drive data into the model and response generators are created to provide accurate responses to the model. The generators directly interface to the FVP at the transaction-level. Separate drivers to control handshaking and signal timing are not required. Stimulus generators can consist of directed tests, random stimulus, or test vectors. Response generators provide the FVP models with application accurate responses to requests and can consist of TLMs from external components or can be developed similar to the FVP models.

5.2.5.7 Creating Interface Monitors

Interface monitors are placed between the TLMs inside the FVP. Interface monitors are commonly used to monitor the signaling on interfaces between blocks, but in the FVP at the transaction-level there is no signaling. Instead the interface monitors observe the transaction interfaces checking higher level protocols and collecting transaction information. The information collected by the interface monitors is useful in debug of the FVP. Interface monitors can identify where a transaction was incorrectly received narrowing the search for the source of a bug. The transaction information collected is also useful in measuring interface coverage within the FVP.

5.2.5.8 Creating Architectural checkers

Architectural checkers monitor the FVP models to verify the correct operation of the FVP. The transaction-level FVP contains little low-level implementation-specific information that needs to be verified. Instead the FVP needs to be verified to meet the specified architectural functional and performance requirements. Architectural checkers are used to verify each of the specified requirements is met. Architectural checks can consist of performance monitors such as latency checks or bandwidth measurements. Architectural checks can consist of comparisons to behavioral models already developed by the architects. Architectural checks can also consist of comparisons to algorithms or datapath models developed in signal processing workbenches.

5.3   Testbench Components

Table 5-2 lists and describes the verification component types used to construct testbenches. Any testbench is possible using these "building blocks." The white boxes with IN, OUT, or I/O (inout) represent ports that communicate transactions.

Table 5-2. Testbench Components

Configuration Master (Top Level)	This is the "top level" test. This block will instantiate all of the blocks described below to construct the verification testbench. Although it has direct access to the verification components it instantiates, for portability reasons it should configure the components their input (target) ports.
Generator	The Generator produces transactions and sends them to one or more output port(s). It is a source of transactions. The Generator can be a test, or it can be a stand-alone module that generates data independently from the test.
Pipelined Generator	Like the Generator, the Pipelined Generator produces transactions. Unlike the Generator, its output is a function of both its internal state and transaction data consumed from one or more input ports. It is typically used to model network protocol stacks.
Transaction Monitor (Response Checker)	The Transactor Monitor is characterized as having one or more transaction input ports. It is a termination point for a transaction; that is, it consumes transactions. Response checkers and scoreboards are examples of Transaction Monitors.
Master Transactor	The Master Transactor is an *abstraction adaptor* that initiates signal-level bus-specific transfers. It consists primarily of a finite state machine that accepts *transaction requests* from the transaction interface and initiates a signal-based protocol to execute the transaction at the signal level. Simulation time advances as this occurs.90

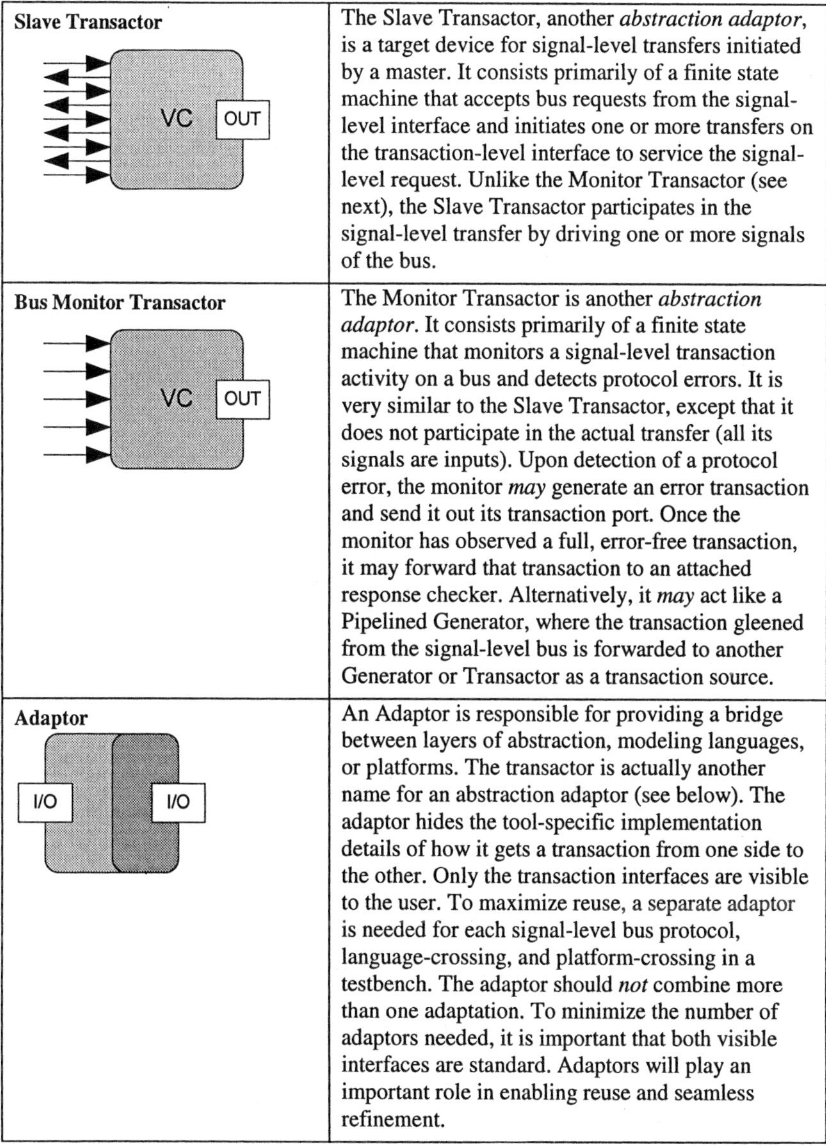

Slave Transactor	The Slave Transactor, another *abstraction adaptor*, is a target device for signal-level transfers initiated by a master. It consists primarily of a finite state machine that accepts bus requests from the signal-level interface and initiates one or more transfers on the transaction-level interface to service the signal-level request. Unlike the Monitor Transactor (see next), the Slave Transactor participates in the signal-level transfer by driving one or more signals of the bus.
Bus Monitor Transactor	The Monitor Transactor is another *abstraction adaptor*. It consists primarily of a finite state machine that monitors a signal-level transaction activity on a bus and detects protocol errors. It is very similar to the Slave Transactor, except that it does not participate in the actual transfer (all its signals are inputs). Upon detection of a protocol error, the monitor *may* generate an error transaction and send it out its transaction port. Once the monitor has observed a full, error-free transaction, it may forward that transaction to an attached response checker. Alternatively, it *may* act like a Pipelined Generator, where the transaction gleened from the signal-level bus is forwarded to another Generator or Transactor as a transaction source.
Adaptor	An Adaptor is responsible for providing a bridge between layers of abstraction, modeling languages, or platforms. The transactor is actually another name for an abstraction adaptor (see below). The adaptor hides the tool-specific implementation details of how it gets a transaction from one side to the other. Only the transaction interfaces are visible to the user. To maximize reuse, a separate adaptor is needed for each signal-level bus protocol, language-crossing, and platform-crossing in a testbench. The adaptor should *not* combine more than one adaptation. To minimize the number of adaptors needed, it is important that both visible interfaces are standard. Adaptors will play an important role in enabling reuse and seamless refinement.

5.3.1 Verification Communication Modes

An interface defines the calling mechanism for communication, and the channel provides the actual transport mechanism over which communication takes place. To support concurrent processing and independent data and control flows, the interface must support several modes of transport.

Figure 5-18 shows the possible modes of communication between two components. All communication requires a Producer and a Consumer, and each of these can be either an active initiator or a passive target. Control flows from Initiator to Target, while data flows from Producer to Consumer.

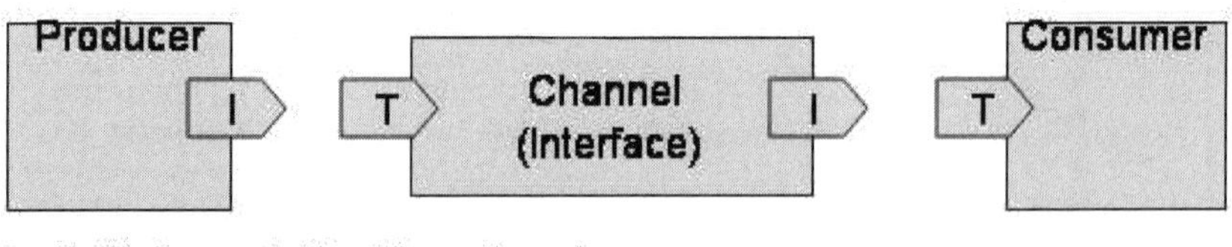

Figure 5-18.

By convention, transaction data always flows left to right. In the above figure 5-18, the Producer is initiating transaction and the Consumer is the target. So, in this case, control flow is occurring left to right as well. If, on the other hand, the Consumer initiates communication by requesting new transactions of the Producer (unsolicited requests for data is termed "pulling"), the Consumer is the Initiator, and control flow would be occurring right to left, opposite the direction of data flow.

Table 5-3, summarizes the possible combinations of Producer/Consumer and Initiator/Target pairings and the implications on how each could communicate.

Table 5-3.

Producer	Consumer	Communication Mode(s) Supported
Producer Initiator	Consumer Target	The Producer pushes transactions to the Consumer, which is otherwise idle. Depending on what process type, if any, is in control of the Consumer, the Channel (think sc_signal) would provide the synchronization services to "wake up" the Consumer on a Producer-Initiated event. If the Consumer has no processes, then there must be a function for the Channel to call in order to pass the transaction on. The function call is the fastest mechanism, as it is the most direct way to execute a transaction.
Producer Target	Consumer Initiator	The Consumer pulls (requests) new transactions from the Producer, which is otherwise idle. The Producer may produce transactions within a thread process, method process, or simple function call. If anything other than a function call, the Channel must notify the Producer thread or method process to "wake up" and produce the requested transaction.
Producer Target	Consumer Target	Both the Producer and Target are not Initiators. To prevent an otherwise deadlock condition, the Channel can act as Initiator for both Producer and Consumer. It will pull a transaction from the Producer and then push the result to the Consumer. Such a mechanism requires the the Channel to have its own thread process.
Producer Initiator	Consumer Initiator	Both the Producer and Consumer are Initiators. To facilitate communication between them, the Channel acts as the Target for both Producer and Consumer. The Producer pushes transactions to the Channel, which stores them temporarily in a fifo. At its own pace, the Consumer will pull transactions from the Channel. The Producer may push transactions in non-blocking or blocking fashion. In the latter case, it may decide to wait until the transaction it has just pushed is retreived from the fifo by the Consumer, until the transaction has actually begun to be processed, or until the transaction has completed execution. The latter scenario is the typical programmer's view for test writing: perform a write transaction and don't return until the write is completed. A subsequent read to the same address would therefore retreive what was just written.

Components may act as both a Consumer and Producer, as in the Pipelined Generator described earlier. The Consumer, for example, upon receiving a transaction from the Producer, may modify it a bit or encapsulate it with some header information, and then send the new or modified transaction out an initiator port for further processing by the next component in line. In this case, the component acts as a Consumer on the input port and Producer on its output port. Multiple components, which by themselves are simple, modular building blocks, can be chained together in this fashion to produce complex stimulus. Simple components such as this also promote reuse, as they could be combined in different ways to produce different stimulus without having to develop another component.

5.3.1.1 SC_THREAD, SC_METHOD, and Functions

The component developer (and test writer, as tests are, in essence, components) may choose whichever programming semantic is appropriate for the function they wish to perform. In many cases, the choice is more or less arbitrary. In general, processing transactions via function call, event-based SC_METHOD invocations, and SC_THREAD context switching provide decreasing performance in exchange for perhaps more intuitive program control.

The SC_THREAD process (and any functions it calls) can suspend and resume at the place it suspended, and it can do so any number of times within the body of the process. It is analogous to Verilog behavioral code, in which statements can be executed sequentially with '@' or '#' delays interspersed between them. Simulation time is allowed to advance during these delays. Software programming is much like this, too, where reads and writes and printf statements and the like all execute sequentially and in blocking fashion, taking a finite amount of "wall clock" time to execute.

The SC_METHOD process is executed start to finish with each invocation and in zero simulation time. This requires the component to manage state information across multiple invocations (usually, the state information is contained as the component class's data members), as simulation time advances between SC_METHOD calls. This mode of processing is analogous to the Verilog 'always' block without any delay statements allowed.

The function call, while fastest, has the most strings attached. It must execute in zero time if the Initiator component called it from within an SC_METHOD process. However, the function may execute wait statements if the calling process is an SC_THREAD. The function must query the SystemC kernel, or the transaction must convey this information about the caller, before attempting to wait. To be safe and simple, the function call is typically written to execute in zero-time. Complete, fully functional testbenches may be written with just function calls, with no additional threads than those in the HDL simulator. However, in such use-models tests are no longer sequentially running "masters". Rather, they are slaves to requests by the transactors for the next transaction to process.

Figure 5-19 below illustrates how a target component can be implemented to support all three methods of process control without much additional code. Both the SC_THREAD and SC_METHOD eventually call the function to process the transaction. Of course, if the function is called via the SC_METHOD process, the function may not execute wait.

```
class component : public uvm_component_t

SC_CTOR(component) {
  SC_THREAD(my_thread);
  SC_METHOD(my_method);
  controller.register("typename",execute);
}

my_thread() {
  while(1) {
   initiator_port.wait_not_empty();
   initiator_port.pull_wait(transaction);
   execute( transaction);
  }
}

my_method() {
  initiator_port.trigger_when_not_empty();
  initiator_port.pull_no_wait(transaction);
  execute( transaction);
}

execute( transaction& t) {
  add_header( t);
  initiator_port2.push_no_wait( t);
  // note: can't use 'push_wait' calls if
  // current process is SC_METHOD
}
```

Figure 5-19. SC_THREAD vs. SC_METHOD vs. function call

5.3.2 Using Transactions in testbenches

Let us now take a closer look at how testbench components are used in a testbench:

Tests of digital hardware that drive and read the signal-level interface of the DUV are difficult to write. Using traditional signal-level verification techniques, the verification engineer is responsible for determining a set of verification vectors that adequately test the functionality of the design. Vectors are an explicit indication of the sequence of transitions of every external signal of the design. Many different scenarios need to be verified. Many portions of the DUV's state space need to be explored. Therefore, truly effective verification vectors are difficult to create. In addition, verification vectors are difficult to maintain when the design itself is still under development. Each time a bug is fixed in the design, the overall timing or sequencing of each test scenario may change. Any change then, necessitates the modification and re-verification of the vector suite itself. Overall, the detail that is required to create good test vectors burdens the test writer who is more interested in questions such as "Is the read data correct?" rather than questions such as "Did I get an acknowledgement on exactly cycle 12?."

In contrast to the test vector approach, transaction-based verification raises verification engineers' test creation efforts to the transaction level thereby increasing their productivity. The transaction level is the level at which users think about the design, so it is the level at which you can verify the design most effectively.

5.3.3 Characteristics of Transactions

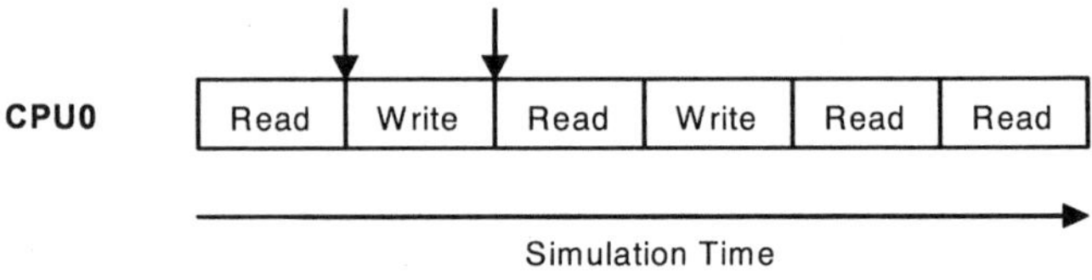

Figure 5-20. Transaction Sequence in a Stream

To aid in debugging the system, you can associate pieces of information called attributes with each transaction. For example, you can show the address and data as attributes of a read transaction. Figure 5-21 shows two transactions, each with three attributes.The topmost attribute is the label, which is a short, user-defined string that identifies the transaction. Below the label are two additional attributes, addr and data.

Read	Write	Read	Label
Addr: 0xffff428a	Addr: 0xffff4200	Addr: 0xffff4280	Addr attribute
Data: 0x07f264b9	Data: 0x01213440	Data: 0x234082	Data attribute

Figure 5-21. Transactions with Attributes

You can use any meaningful attribute on a transaction. There is no limit to the number of attributes that you can associate with a transaction.

5.3.4 Hierarchical Transactions

Transactions sometimes occur in phases. The burst read shown in figure 5-22 is an example of such a transaction. The burst read operation begins with an initial phase, Setup, followed by four single-byte reads. The subordinate phases are called child transactions. Their parent is the top-level burst read. In verifying a design, sometimes you need to consider only the parent transaction and sometimes you need to examine the finer granularity of the child transactions.

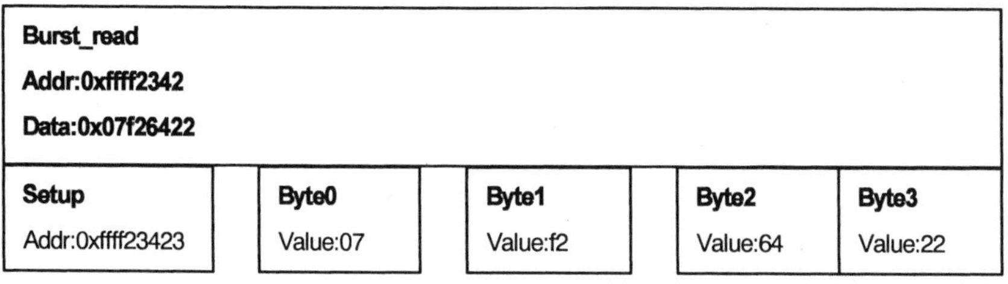

Figure 5-22. Hierarchical Transaction

Child transactions occur in the same stream as their parent. A child transaction can begin at or after the same time as its parent begins, and must end before or at the same time as its parent ends. That is, child transactions cannot extend outside of the time bounds of their parent. There is no limit to the number of levels of hierarchy that you can create.

You can associate attributes with children. The attributes of a parent are considered to be the attributes of the child if the child does not specifically set them. In the example in Figure 5-22, the children of the read transaction also have addr and data as attributes.

5.3.5 Related Transactions in multiple streams

Typically, you record several different types of transactions in a single stream. Many protocols, such as protocols for pipelined processor buses, use multiple overlapping operations. Because NCSystemC supports overlapping transactions in a single stream, you can implement these protocols by using one or more interacting streams. Such protocols represent two independent, but related, interfaces or streams within a single transactor. These interfaces operate concurrently and independently, but data and control information moves between them.

TBV supports the independent cause-and-effect relationships between transactions that result from the expected operation of the design itself. System interfaces can interact in the following ways:

- Closely - An example is when the address and data phase interfaces of a pipelined processor bus protocol transfer data between a bus master and a bus slave.

- Loosely - For example, in a design that bridges traffic from the processor bus to an I/O expansion bus, the activity on the processor interface might cause some related activity to occur on the I/O interface at some later time.

A transaction that causes another transaction to occur is the predecessor of the resulting transaction. The resulting transaction is the successor of the initiating transaction. For example, you might be testing a system where the processor transactor controls both an address phase interface and a data phase interface. The test issues a read request to the processor transactor, which initiates a transaction in the address phase stream. Some time later, when the system returns the read data, the related data phase transaction occurs in the data phase stream as shown in Figure 5-23:

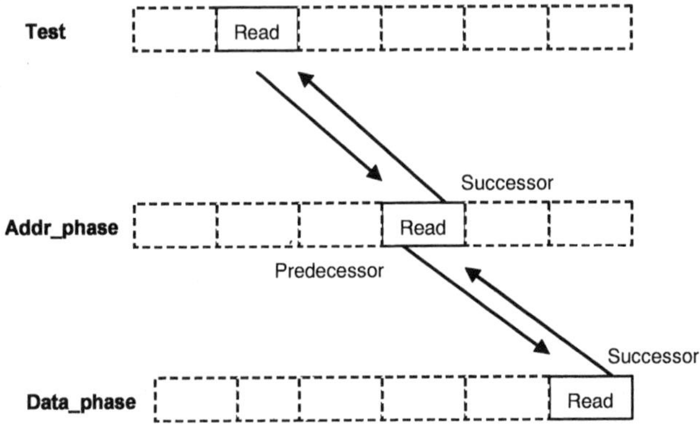

Figure 5-23. Multiple Interacting Streams

NCSystemC can record a relationship in a simulation database. It can be viewed later using a viewer that supports transactions.To keep track of relationships between transactions that occur in different streams, you can link transactions together. A typical relationship is the predecessor/successor relationship. NCSystemC uses transaction handles to manipulate transactions, and strings for describing a relationship.

5.3.5.1 Implementation of Transactions in SCV

A transaction object can be created in the C++ language by defining a transaction class. A transaction class should contain the following elements:

- All the relevant information that describes the transaction. For example, the transaction type and the transaction attributes such as address and data.

- Constructors to create a default type of transaction and specific transactions

- Utility functions that operate on transactions. For example, a function to display the contents of a transaction to output that information to a log file or to the screen of a interactive debugger window.

- A destructor if the transaction class needs to clean up dynamically allocated memory or do other house keeping like closing files, and so on.

the following listing is an example transaction class for the AMBA APB interface:

```
enum apb_transaction_type_t {APB_TX_IDLE, APB_TX_READ, APB_TX_WRITE};

struct apb_arg_blk_t
{
enum apb_transaction_type_t type;
sc_lv<num_psel>      sel;
sc_lv<data_width>    data;
sc_lv<addr_width>    addr;

apb_arg_blk_t() : type(APB_TX_IDLE), sel(0), data(0), addr(0) {};
apb_arg_blk_t(enum apb_transaction_type_t t,
          sc_lv<num_psel> s,
        sc_lv<data_width> d,
        sc_lv<addr_width> a) : type(t), sel(s), data(d), addr(a) {};
~apb_arg_blk_t() {// no dynamic data};
show() {
 cout << type << " " << sel << " " << data << " " << addr << endl;
}
};
```

The APB protocol used in this example contains a signal-level interface that includes address and data wires as well as wires for a clock, device select, reset and device enable. This transaction class is an abstraction of that signal-level interface, so it does not contain the clock and the signals needed for device handshaking. This transaction class contains only the address, data and type of a transaction (read, write, idle) for the simple bus APB. The size of address and data are parameterized using templates. The width values for these variables can be defined outside of the class definition, at a higher level so that multiple objects in the simulation can share the same bit width parameters. Two constructors are shown:

1. a default constructor that initializes the transaction to IDLE and

2. a three-argument constructor that uses the arguments to initialize the transaction

.The show() method is provided for use in the source level debugger to display the values of the class data members.

If you want to randomize or record an instance of a transaction object your need to create an extension class that has methods for these, and other operations. A wizard program is provided with the NCSystemC to produce automatically a class extension header file from a transaction class definition header file.

A transaction class may inherit from an abstract base class that defines a standard programmer's interface to transaction objects. This abstract base class obligates transaction class creators to provide implementations of the member functions in the base class.

An example transaction base class might look like:

```
class uvm_tx_base_t {
  virtual ~uvm_tx_base_t() = 0;
  void show() = 0;
  void record() = 0;
  const char * convert_to_bytes() = 0;
  void convert_from_bytes(const char *) = 0;
  // etc..
};
```

The conversion methods shown might be useful for some transactions to create serial byte streams from the transaction or to convert serial byte streams into transactions. This kind of function is often used in network communication protocols where one network protocol layer's transaction is simply the byte stream payload of a lower layer's transaction.

5.3.5.2 Transactor implementation in SCV

In transaction-Based Verification the test does not directly connect to the design under verification. Instead, transactors represent the interface between the test and the design. The test sends transaction requests to the transactors. The transactors translate the transactions into detailed, signal-level activity. Likewise transactors observe signal-level interfaces and encode the activity seen there as transactions which can be reported back to the test. This strategy lets you develop tests at a higher level of abstraction, so that you can easily create and maintain the many variations required to thoroughly test a design's interaction with its environment.

Transactors have the following qualities:

- A transactor is a mechanism for the modular design of test benches that uses information hiding. In general, information hiding means that "each module hides some design decision from the rest of the system". A transactor hides a great deal of design information about the state machine that communicates with a signal-level interface. The user of the transactor does not need to know much about the protocol that the hardware interface implements. The user of the transactor only needs to know about the transaction-level interface that the transactor presents for the purpose of writing hardware verification tests. The transaction-level interface that the transactor user sees is abstract and therefore easy to learn and use. The signal-level interface through which the transactor communicates with the DUV is hidden from the transactor user.

- A transactor fits in a hierarchical structure for hardware verification starting with the DUV at the bottom of the hierarchy, moving up through transactors to test components, and finally to the master controller of the simulation.

- Composing complex systems in a hierarchical fashion is a technique pioneered in operating system design and later adopted for the design of network protocols. This approach promotes reuse when there are well-defined interfaces between the layers. Then the implementation of one layer can be replaced with another implementation or a particular implementation of a layer can be reused in another hierarchy.

- Finally, a transactor is an object-oriented class. "The class concept encapsulates both the substance, the representation of values and the operations" that form the transactor.

There are three different types of transactors depending on their relationship to the DUV and the testbench:

- Master transactors — Connect a verification test program to a signal-level interface, which is part of a device under verification (DUV). The master transactor takes a transaction from a stimulus generator and converts this information into the signal-level protocol of the interface.

- Slave transactors — Connect to a DUV interface and an abstract model of a target device being emulated. A slave transactor recognizes transactions on a signal-level interface and responds to these transactions, under the control of the verification test, by emulating a physical device, as if the device were actually connected to the interface.

- Monitor transactors — Connect to a DUV interface and recognizes transactions. Unlike a slave transactor, a monitor transactor only reports what it sees to a higher level verification component (such as a response checker).

Other transaction-based verification components that work with transactors are:

- Stimulus generators — Generate transactions that are driven into the DUV through transactors.

- Response checkers – Connect to monitor transactors and determine if DUV behavior is what is expected. They report errors such as protocol violations by the DUV or transaction outputs from the DUV that are incorrect.

- Reference models – Are abstract models of DUV behavior that can be used by response checkers to determine what the correct response of the DUV should be. A reference model is usually implemented with entirely transaction-level interfaces, not signal-level interfaces and is often called a transaction-level model (TLM.)

The following diagram in figure 5-24 illustrates the relationship of these components:

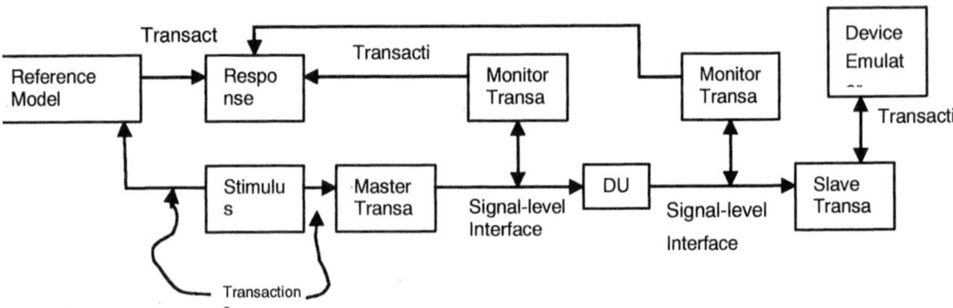

Figure 5-24. TBV Testbench Components

5.3.5.3 Transactor Architecture

A transactor connects to a device under verification through a signal-level interface. The transactor connects to a higher level component (such as a stimulus generator in the case of a master transactor) through a transaction-level interface. The
signal-level interface is based on SystemC sc_port objects that map to signals in the
DUV. The transaction level interface from the transactor to a higher level component is based on call backs of methods in a SystemC sc_interface object.

Master Transactors

The callback methods for a master transactor can implement a stimulus generator
directly (Pull interface) or they can fetch transactions that are placed in a FIFO
queue by a stimulus generator (Push interface.) The following diagram in figure 5-
25 illustrates this arrangement:

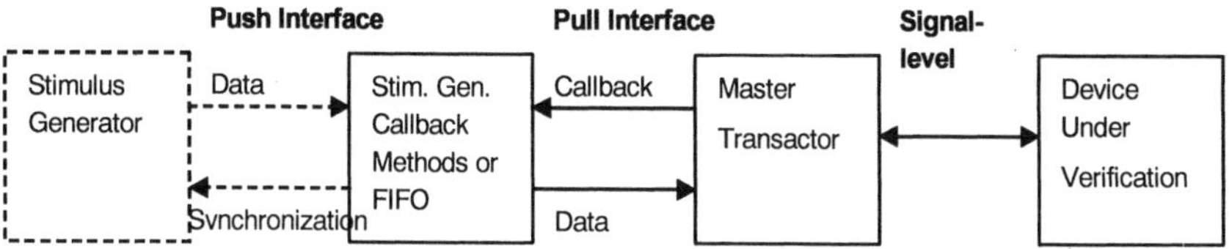

Figure 5-25. Push-Pull Transactor Interface

When the Pull interface is used, the stimulus generator is implemented as methods
called by the master transactor. The Push interface is layered on top of the Pull
interface. A FIFO queue is placed between the stimulus generator and the master
transactor's callback methods. The components used for the Push interface are
shown in dashed lines to indicate that it is an option that can be attached to a Pull
interface.

The Pull Interface

The most efficient way for a transactor to communicate with other verification
components is through callback functions. This approach is called the Pull interface
because it allows the transactor to pull data from another component.

A transactor with a Pull (or callback) interface is able to transfer the flow of program control to another verification component without the need to create a new

thread or incur the overhead of a thread context switch. Another benefit to the Pull approach is that data exchange with the transactor occurs as close to the time the data is used as possible. A minimum amount of data queuing or perhaps none at all is required. This makes it possible to build highly reactive test benches.

Consider the situation where a master transactor pulls data from a stimulus generator by way of callbacks. The transactor calls the stimulus generator just before the data is needed. At that instant the stimulus generator can determine the current state of all the components of the simulation that it is concerned with. Then the stimulus generator can decide what stimulus to send to the transactor based on all the late-breaking activity that it may be monitoring.

For example, in a protocol like AMBA AHB where a slave can issue a split response to a master it is difficult for the stimulus generator to anticipate when splits will occur. If the stimulus generator wants to do something special to stress the system when it sees a split from a particular slave, it can do so if the stimulus generator is communicating with the master transactor using a Pull interface.

A Pull interface consists of five basic parts:

1. A SystemC interface definition. For example:

```
template <typename TX_T>
class uvm_generator_if_t : virtual public sc_interface {
public:
  virtual void get_data(TX_T &) = 0;
};
```

UVM transaction interface classes are template classes. The interface member functions take as their single argument a user-defined class (of type TX_T in the above example.) This class type of the member function argument will change for a different type of transactor. For example a PCI transactor will have a different transaction class than an AMBA AHB transactor. However, both sets of transactors will share the UVM interface class, so they will both have get_data(TX_T&) member functions. Using templates in this way provides several benefits:

- All transactors share the same basic interface which only varies in the type of the transaction class. This makes the transactors easier to understand and use. Once one type of transactor is used, there is some idea of how another one of another type will work.

- Other components which connect to transactors but which do not care about the type of the transaction class can be reused readily. For example a component that queues, stores or records transactions and therefore doesn't need specific knowledge of the transaction type can be completely reused.

2. The declaration of a SystemC port in the transactor. Consider the following example:

```
class apb_master_t : public sc_module {
public:
  sc_port<uvm_generator_if_t<apb_arg_blk_t>,1> generator;
};
```

The sc_port connects an APB master transactor to a stimulus generator through a transaction-level interface.

The sc_port has a template type that is the name of the interface class. In this case the template type is: uvm_generator_if_t. That type in turn has its own template parameter which specifies the template class handled by this generator, e.g. apb_arg_blk_t.

The second argument in the port's template is the number of objects attached to the port. A value of 0 means any number of objects. For a stimulus generator you will usually have just one connection. A monitor transactor may have any number of connections.

3. The definition of a class that inherits the SystemC interface definition (as defined in item 1 above.) This class contains the callback function (get_data()), which is a virtual function in the interface class.

```
class apb_master_generator_t : public uvm_generator_if_t<apb_arg_blk_t> {
public:
  apb_master_generator_t() {}
  ~apb_master_generator_t() {}
  virtual void get_data(apb_arg_blk_t &arg);
};
```

4. The definition of the interface member function, where stimulus data is created in the case of a stimulus generator:

```
void apb_master_generator_t::get_data(apb_arg_blk_t &arg) {
scv_smart_ptr<apb_arg_blk_t> arg_sp;
arg_sp->next();
arg = arg_sp->read();
return;
  }
```

5. The declaration of the interface class, usually in the same module that declares (in other words, instantiates) the transactor:

```
class sctop_t : public sc_module {
```

```
public:
  apb_master_generator_t master_gen;
};
```

Signal-Level Interfaces to the DUV

The DUV connected to a transactor is usually an RTL model with an interface made up of bits or bit-vectors. Often the interface to a DUV is a standard bus like AMBA AHB, PCI, or USB, and so on. The examples that follow begin by using a very simple bus interface: an AMBA APB interface. Later examples use more complex interfaces to illustrate other techniques.

The DUV that a transactor connects to may be written in SystemC or a traditional HDL like Verilog or VHDL. At the time of this writing the vast majority of RTL designs are done using Verilog or VHDL. Much of the discussion here focuses on Verilog DUVs.

The upper layers of a transactor and more abstract UVM components like stimulus generators and response monitors are independent of the language in which the DUV is written. The DUV could be written in Verilog, VHDL, SystemC, or perhaps some other HDL. This division of the verification components into layers is a means of promoting reuse through abstraction.

The following example uses NC-SystemC to connect a SystemC transactor to HDL by creating a shell HDL module for AMBA APB using Verilog.

Example 1-1. Connecting a SystemC Transactor to HDL Using a Shell HDL Module

```
// apb_master.v
//
// A Verilog wrapper around a SystemC APB master transactor.
// Used to instantiate the master in an HDL simulation model.
// The master is defined in the C++ file apb_master.{h,cpp}
//
// Note module name must match exactly name of sc_module class
// in SystemC.
//
module apb_master_t(PENABLE,
PADDR,
PWRITE,
PWDATA,
PRESET,
PCLK,
```

```
                       PSEL,
                       PRDATA)
//
// Note that the foreign attribute string value must be "SystemC".
//
(* integer foreign   = "SystemC";
 *);
//
// Note that port names must match exactly port names as they appear in
// sc_module class in SystemC; they must also match in order, mode and
// type.
//
  parameter addr_width_p = 32,
     data_width_p = 32,
         num_psel_p  = 4;
  output  PENABLE;
  output [addr_width_p-1:0] PADDR;
  output               PWRITE;
  output [data_width_p-1:0] PWDATA;
  input  PRESET;
  input          PCLK;
  output [num_psel_p-1:0]   PSEL;
  input [data_width_p-1:0]  PRDATA;
  endmodule
```

The Verilog module above in Example 1-1 is the wrapper that allows NC-SystemC to instantiate a SystemC module. There is one more piece of Verilog that is required to complete the instantiation. The wrapper module itself needs to be instantiated in a top-level Verilog module as shown in the following example. With NC--SystemC a Verilog module can instantiate a SystemC model using the same syntax used to instantiate a Verilog instance. The OVI attribute "SystemC" in the HDL wrapper indicates that what is instantiated is a SystemC module.

Example 1-2. Instantiating a Wrapper in a Top-Level Module

```
        module top;
        parameter addr_width_p = 32,
           data_width_p = 32,
           num_psel_p  = 4;
        wire          PENABLE, PWRITE;
        wire[num_psel_p:0]  PSEL;
        wire[addr_width_p:0] PADDR;
        wire[data_width_p:0] PWDATA;
        reg            PCLK, PRESET;
        wire[data_width_p:0] PRDATA;
        //
        // Instantiate the SystemC HDL Shell
```

```
//
apb_master_t #(addr_width_p, data_width_p, num_psel_p)
apb_master_1(PENABLE, PADDR, PWRITE, PWDATA, PRESET,
PCLK, PSEL, PRDATA);
endmodule
```

It is during the instantiation of these transactor shells that the default parameter values can be overridden to match the characteristics of the DUV that is also instantiated at this level. Note that the wire and reg objects declared in the Verilog code above are the arguments passed to the transactor shell apb_master_1. These represent the signal level interface of the transactor, or in other words, its connection to the DUV.

A number of steps must be taken in the SystemC code of a transactor to enable its connection to an HDL model. You must create a class which represents the transactor. The class declaration includes declarations of all the interface signals that connect to the DUV, making sure that their width and type match the width and type specified in the HDL.

For example, a Verilog output of one bit should map to a SystemC type of sc_out<sc_logic> as PENABLE does above. Likewise a Verilog input bit vector (like PRDATA) should map to a SystemC type of sc_in<sc_lv<data_width> >. The complete declaration of the APB master transactor in SystemC looks like the following example:

Example 1-3. APB Master Transactor In SystemC

```
class apb_master_t : public sc_module {
public:
sc_out<sc_logic>          PENABLE;
sc_out<sc_lv<addr_width> > PADDR;
sc_out<sc_logic>          PWRITE;
sc_out<sc_lv<data_width> > PWDATA;
sc_in<sc_logic>          PRESET;
sc_in<sc_logic>          PCLK;
sc_out<sc_lv<num_psel> >   PSEL;
sc_in<sc_lv<data_width> >  PRDATA;
bool debug;
SC_CTOR(apb_master_t) :
 PENABLE("PENABLE"),
 PADDR("PADDR"),
 PWRITE("PWRITE"),
```

```
    PWDATA("PWDATA"),
    PRESET("PRESET"),
    PCLK("PCLK"),
    PSEL("PSEL"),
    PRDATA("PRDATA"),
    debug(1) {
      SC_METHOD(run);
    sensitive << PCLK.pos();
    }
    void run();
    ~apb_master_t() {}
    };
```

Note in the case of the APB interface that an SC_METHOD is used which is sensitive to the positive edge of PCLK. The APB master does not need to wait on any signals once it is triggered. If it did, we would need to use SC_THREAD instead of SC_METHOD. It is a good idea to name each signal in the class constructor, for example PENABLE("PENABLE"). This may aid in debugging the transactor. If these data members are not given any name, SystemC will and it is preferrable to give them your own names.

The C++ code that defines the transactor should include at file-scope level (which is outside a function or class definition) the NCSystemC provided macro which takes the name of the class which represents the transactor as illustrated in the following line:

```
SC_MODULE_EXPORT(apb_master_t);
```

Summary of Steps Needed To Connect An HDL DUV To A SystemC Transactor

In summary, the following list are the steps necessary for you to perform in order to connect a SystemC transactor to an HDL DUV using NCSystemC:

- Create an HDL wrapper for the transactor which contains:
    - The SystemC foreign attribute.
    - Parameterized declarations of the interface signals.
    - A module with the same name as the transactor's C++ class (apb_master_t) and with an argument list that includes the signals in the interface.
- Instantiate the HDL wrapper for the transactor which takes as its argument list the signals in the interface.

- Create a SystemC definition of the transactor that includes definitions of the interface signals that match the type and width of the HDL signals. Specify the member function that will contain the transactor's state machine.

- Include the SC_MODULE_EXPORT macro in the code that defines the state machine method.

The definition of a SystemC transactor and its Verilog wrapper is shown schematically:

```verilog
Module top;              //module that instantiates the transactor wrapper and
   wire PENABLE;         //connects to the DUV signals
   reg PCLK;
   ...
   apb_master_t apb_master(PENABLE,
                           PCLK,
                           ...
   ...
endmodule
```

```verilog
Module apb_master_t(PENABLE,       //transactor wrapper
                    PCLK,
                    ...);
(* integer foreign = "SystemC";*)
  output PENABLE;
  input CLK;
  ...
endmodule
```

Mapped ports connecting System C
transactor to Verilog DUV

```cpp
class apb_master_t : public sc_module {
  public:
    sc_out<sc_logic> PENABLE;
    sc_in<sc_logic> PCLK;
    ...
    SC_CTOR(apb_master_t) :
      PENABLE("PENABLE"),
      PCLK("PCLK"),
      ...
      debug(1) {
        SC_METHOD(run);
        sensitive <<PCLK.pos();
}
SC_MODULE_EXPORT(apb_master_t)

void apb_master_t::run() {
...    // state machine goes here
}
```

State Machines in SystemC

A state machine written in SystemC looks similar to one written in an HDL like Verilog. State machines are frequently implemented with one or more case statements as their central feature. For example, the following listing is a state machine for the master transactor of an AMBA APB interface:

```
//
// This is the main loop of the master. It runs on the positive edge of PCLK.
//
void apb_master_t::run() {
  static apb_arg_blk_t arg;
  static enum apb_states_t current_state, next_state = APB_STATE_RESET;
  current_state = next_state;
  switch(current_state) {
  case APB_STATE_RESET:
    PENABLE = SC_LOGIC_0;
    PSEL  = 0;
    if (PRESET == 0) {
      next_state = APB_STATE_IDLE;
    }
    break;
  case APB_STATE_IDLE:
    get_and_send_tx(arg, next_state);
    break;

  case APB_STATE_SETUP:
    PENABLE = SC_LOGIC_1;
    next_state = APB_STATE_ENABLE;
    break;

  case APB_STATE_ENABLE:
    PENABLE = SC_LOGIC_0;
    if (arg.type == APB_TX_READ) {
      arg.data = PRDATA;
    }
    for (int i = 0; i < reporter.size(); i++) {
      reporter[i]->report_data(arg);
    }
    get_and_send_tx(arg, next_state);
    break;
  }
  return;
}

void apb_master_t::get_and_send_tx(apb_arg_blk_t& arg, apb_states_t &next_state)
{
  generator[0]->get_data(arg);
```

```
   if (arg.type != APB_TX_IDLE) {
     PSEL  = arg.sel;
     PADDR = arg.addr;
     if (arg.type == APB_TX_WRITE) {
     PWRITE = SC_LOGIC_1;
     PWDATA = arg.data;
   }
   else {
     PWRITE = SC_LOGIC_0;
   }
   next_state = APB_STATE_SETUP;
   }
   else {
    next_state = APB_STATE_IDLE;
   }
  }
  void apb_master_t::debug_on() { debug = 1;}
  void apb_master_t::debug_off() { debug = 0;}
```

State Machines in HDL

A transactor state machine written in HDL will likely simulate faster than one written in SystemC because there will be no signal-level traffic between SystemC and the HDL. The SystemC to HDL communication will be at the transaction level which should occur less frequently than the signal-level traffic. In addition an HDL state machine can be synthesized and executed in an emulator for the maximum in simulation performance.

A state machine written in Verilog looks similar to one written in C++. Usually a case statement is used in the same way as with C++. A big difference between Verilog and C++ is the absence of data structures in Verilog. This means that the transaction-level interface to the transactor must use scalar Verilog types (e.g. reg or wire.)

Slave Transactors

Slave transactors are largely passive components that respond to activity on a signal-level interface. The Pull or callback interface method works very well for them. Whenever a slave transactors state machine needs to get data from the component that is emulating the behavior of the device the slave represents, the slave transactor calls the callback function that provides the required data.

The following example is based on the AMBA APB bus. This bus is so simple that the state machine for a slave can be implemented without using a case statement. Only a set of nested if statements is necessary:

```
//
//This is the main loop of the slave. It runs whenever there is a change in the level of PENABLE.
//
void apb_slave_t::run() {
  static apb_arg_blk_t arg;
  if (PSEL == 1) {
   if (PENABLE == 1) {
     arg.type = (PWRITE == SC_LOGIC_1) ? APB_TX_WRITE : APB_TX_READ;
     arg.addr = PADDR;
     if (PWRITE == 0) {
       generator[0]->get_data(arg);
   PRDATA   = arg.data;
     }
     else {
   arg.data = PWDATA;
   generator[0]->get_data(arg);
     }
   }
   else {
     PRDATA = all_z;
   }
  }
  return;
}
```

This slave has an interface to a SystemC component that emulates a hardware device. The interface operates through the called back function generator[0]->get_data(arg). This function is connected to the slave transactor using a Pull interface as described above. In this example the slave connects to only one device, which is a random access memory (RAM).

The single port in the generator object that contains the callback function get_data is referenced with the array subscript [0]. If a slave needed to be connected to more than one emulated device, to model a hardware structure where more than one peripheral shared a single bus interface, then the slave's state machine could selectively call more than one get_data method by using an array subscript other than 0.

The RAM in this example is modeled with a sparse array so that a potentially large address range can be modeled with a small amount of program memory.

```
void apb_slave_generator_t::get_data(apb_arg_blk_t &arg)
```

```
{
static map<sc_uint<addr_width>, sc_uint<data_width> > array;
  if (arg.type == APB_TX_READ) {
    if (array.find(arg.addr) != array.end()) {
      arg.data = array[arg.addr];
    }
    else {
      cout << "apb slave generator error: read from uninitialized"
      << " array element at address: " << arg.addr << endl;
      arg.data = all_x;
    }
  }
  else if (arg.type == APB_TX_WRITE) {
    array[arg.addr] = arg.data;
  }
  else if (arg.type == APB_TX_IDLE) {
    // nothing to do in this case
  }
  else {
    cout << "apb_slave generator error; invalid transaction type: "
  << arg.type << endl;
  }
}
```

The slave state machine is decoupled from the device that the slave actually represents. In this way the state machine can be reused and attached to a different emulated component, an I/O controller for example, depending on the configuration requirements of the test bench.

The interface to the emulated RAM is the same one that is illustrated in section The Pull Interface above.

```
class uvm_generator_if_t : virtual public sc_interface {
public:
virtual void get_data(apb_arg_blk_t &) = 0;
};
```

Likewise, the RAM emulator is connected to the slave transactor by way of a port, which is declared inside the slave transactor definition:

```
class apb_slave_t : public sc_module {
public:
sc_port<uvm_generator_if_t<apb_arg_blk_t,1> generator;
}
```

Finally, the class containing the emulator function definition is declared, usually in the same place (such as in a high-level module in the test bench) as the class for other stimulus generators. For example:

```
class sctop_t : public sc_module {
public:
apb_master_generator_t master_gen;  // stimulus gen. for master
apb_slave_generator_t slave_gen;    // dev. emu. for slave
}
```

Monitors

A monitor transactor is entirely passive with respect to the interface it is monitoring. An interface's signals are inputs to the monitor. It never drives an interface signal. A monitor communicates with a test component like a response checker using a call-back method attached to an sc_port. Masters and slaves use a similar mechanism but in their case the data transfer is from the transaction interface to the transactor. In the case of the monitor the data transfer is in the other direction: from the transactor to the transaction interface.

A monitor transactor is a transaction recognizer. To perform this function the monitor usually has a state machine attached to the signal interface. At the point where a transaction has been recognized and all the transaction attributes (e.g. type, address and data) have been seen by the monitor, it calls the call-back function to report the transaction. E.g. the reporting portion of an APB monitor transactor looks like this:

```
//
// apb_monitor.cpp
//
// A monitor transactor for the AMBA APB bus.
// The SC_METHOD run() executes on the positive edge of PCLK
//
#include "apb.h"
#include "apb_monitor.h"
#include "uvm_util.h"
SC_MODULE_EXPORT(apb_monitor_t);
void apb_monitor_t::run() {
  static enum apb_states_t previous_state, current_state = APB_STATE_RESET;
  static apb_arg_blk_t arg;
  previous_state = current_state;
  sc_lv<num_psel> psel = PSEL.read();
  switch(previous_state) {

    ... // multiple cases removed to shorten example

  case APB_STATE_SETUP:
    if (psel == 0) {
cout << "monitor -- error: PSEL deasserted in the ENABLE state" << endl;
    return;
```

```
      }
    else if (uvm_check_for_X(psel)) {
cout << "monitor -- error: PSEL is X in the ENABLE state" << endl;
return;
      }
    else if (uvm_check_for_Z(psel)) {
cout << "monitor -- error: PSEL is Z in the ENABLE state" << endl;
return;
      }
    else if (!uvm_check_for_one_bit_set(psel)) {
cout << "monitor -- error: More than one PSEL bit is set" << endl;
 return;
    }
   if (PENABLE == 1) {
if (debug) cout << "monitor: state is ENABLE" << endl;
  current_state = APB_STATE_ENABLE;
  arg.sel  = psel;
  arg.type = (PWRITE == SC_LOGIC_1) ? APB_TX_WRITE : APB_TX_READ;
  arg.addr = PADDR;
  arg.data = PWDATA;
  for (int i = 0; i < reporter.size(); i++) {
    reporter[i]->report_data(arg);
  }
    }
    else {
cout << "monitor -- error: PENABLE not asserted in the ENABLE state" ;
<< endl;
    }
   break;

   ... // rest of code removed
```

The line:

```
reporter[i]->report_data(arg);
```

is where data is sent from the monitor to other test bench components. Note than any number of components may be registered to receive data from a monitor. That is why this call back is in a for loop and executes once for each receiver that are registered as connection with this monitor.

Instantiation And Connection Of Transactors And Other Components

Transactors are instantiated and connected to the DUV from HDL code that references the HDL modules that are transactor wrappers. E.g. to connect a trio of APB transactors -- a master, slave and monitor you might have this in one of your Verilog modules:

```
module top;
parameter addr_width_p = 32,
   data_width_p = 32,
   num_psel_p  = 1;
wire     PENABLE, PWRITE;
wire[num_psel_p:0] PSEL;
wire[addr_width_p:0] PADDR;
reg      PCLK, PRESET;
wire[data_width_p:0] PRDATA, PWDATA;
//
  // Instantiate the SystemC HDL Shells for master, slave and monitor
  //
  apb_master_t #(addr_width_p,data_width_p,num_psel_p) apb_master_1(PENABLE, PADDR, PWRITE,
PWDATA, PRESET, PCLK, PSEL, PRDATA);
apb_slave_t #(addr_width_p,data_width_p) apb_slave_1(PENABLE, PADDR, PWRITE, PWDATA, PSET,
PCLK, PSEL[0], PRDATA);
apb_monitor_t #(addr_width_p,data_width_p,num_psel_p) apb_monitor_1(PENABLE, PADDR, PWRITE,
PWDATA, PRESET, PCLK, PSEL, PRDATA);
... // rest of code removed
endmodule
```

Components like stimulus generators and device emulators are not connected to HDL and so are not instantiated from HDL. They are instantiated from SystemC in a module that defines the configuration of a test bench. Since these types of components connect to transactors which are instantiated by the HDL simulator it is necessary to get a handle to the transactors by using the SystemC method simcontext()->find_object(const char *). For example to get a handle to the APB master transactor and to attach it to a stimulus generator you would use this approach:

```
void sctop_t::end_of_construction()

{

apb_master_generator_t master_gen;

apb_master_t *master_p = (apb_master_t*) simcontext()->find_object("top.apb_master_1");

master_p->generator(master_gen);

...

}
```

These statements are put in a member function of the class that configures the test bench. In this case that class is called sctop_t. The member function used is a member of the SystemC class sc_module. The function -- named end_of_construction()

-- is called by the SystemC process that constructs modules and their interconnection.

The first statement inside the end_of_construction member function declares a stimulus generator. The second statement gets a handle to the master transactor. The third statement connects the stimulus generator to the master transactor.

CHAPTER 6

Verification Methodology

The increasing complexity of ICs poses challenges to both system design engineers and verification engineers. The industry has responded to this challenge by adopting design reuse strategies to increase the productivity of design processes. Every reusable design block needs to be accompanied by reusable and complete verification suite for thorough intra-block verification. Every design/company have their own verification suite and methodologies based on the requirement and the resources. It is very important that a standard verification methodology is defined that is suitable for different design requirements in order to encourage the reusablilty of verification components. Also, with company's focus on time-to-market and first pass silicon success, it is critical to develop a verification methodology that is independent, reusable and self checking. This chapter discuss functional verification methodology adaptable to any networking design with examples taken from networking reference design.

Topics covered in this chapter are

- Verification approach.
- Functional verification tools.
- Transaction based verification environment.
- System Level verification.
- Block Verification.

- Block TVMs at system level.
- Infrastructure.
- Stimulus/Traffic Generation.
- Rate Monitors.
- Checkers.
- Memory models.

6.1   Introduction

The best verification methodology is the one that allows,

- reusability
- self checking
- is independent of design cycle

There are several verification tools and environments available as discussed in the "Verification Processes" chapter. Verification methodology is dependent on the design requirement and the available resources. There are various high level verification languages available in the market to choose from. The methodology discussed here is based on the usage of C++ as a high level verification language using a transaction based verification approach.

A Self-checking technique is most reliable and allows extensive verification. It does not depend on large pool of reference models or golden reference files, and provides   the verification results as simulation is progressing based on actual and expected responses. Overall verification approach and transaction based verification is discussed in following section.

6.2   Overall Verification approach

Figure 6-1 represents a typical verification flow. Verification approach should be a logical verification flow, moving from the system level through individual block level verification. Design verification planning should start concurrently with the creation of system specifications. System specifications should drive the verification strategy.   Based on the verification plan, a set of transaction verification modules (TVMs) should be written for generators, monitors and checkers. Plug and play approach should be supported which means that even if the RTL block is not ready, It should be possible to run simulations at the system level using these TVMs to help refine and complete the whole system level verification infrastructure. As block RTLs become available they can be replaced with the TVMs which were used before. Later, these TVMs for respective blocks can be used for probing at block interface.

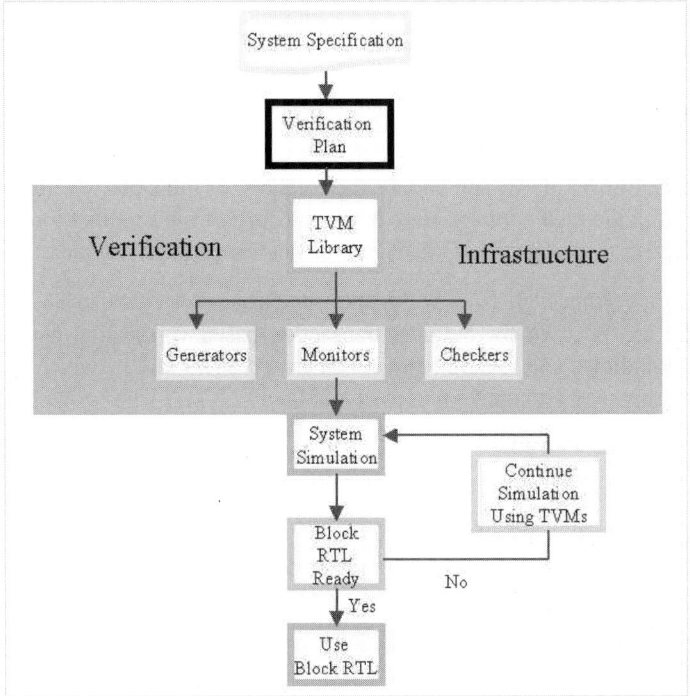

Figure 6-1. Verification flow

The verification infrastructure should be developed so that the same infrastructure of generators, checkers and monitors can be used for block level verification also. Since the environment is independent and reusable, block designers can write stimulus tests based on their test plan using the environment developed for system level verification. Checkers should be written by verification engineers for checking functionality of the chip at the input and output interfaces. Ideally, checkers should be developed for checking individual block functionality during top level simulation, where blocks are tied together at the system level.

While doing system verification and debugging, it should be possible to trigger checking of specific block to be on or off, and expect the output at the block interface in addition to the system output. It adds to the confidence on the blocks behavior since verification engineer writes checkers.

One main advantage of having checkers for all the blocks at top level is that with small modifications they can also be used as models for blocks if block is not ready to be integrated at system level.

6.3   What all tools are needed

It is essential to list down and choose the category of tools that will be used for implementing the methodology. Here is the category of tools required for successful verification. More details of this is in "verification process" chapter.

- Tools supporting high level verification language. For example SCV or Testbuilder can be used for building transaction based verification environment. This tool allows you to define test bench environment in C++ which speeds up test creation and maximizes reuse without the PLI/VHPI knowledge.
- Tools for Lint checking.
- Tool for code coverage.
- Tool for functional coverage.
- Standard Template Library (STL) for implementing data structures of lists, queues, associative arrays etc.
- Tool for Version control. It helps during the development process of verification and allows everyone to share the same verification infrastructure.
- Bug tracking tool for reporting and tracking all the issues. It is a good idea to choose web based bug tracking tool.
- Memory model for functional simulation.

Here is the example set used for reference design:

Function	Tool Name
Transaction Based Verification	TestBuilder
Advanced Analysis (Lint Checking , Assertion Basedand Code coverage)	Hardware Lint Checking (HAL), 0-In, SureCov
Simulation	NCSim
Functional Coverage	Transaction Explorer
Compiler	G++ (2.9.5)
Class Library	STL
Revision Control	CVS
Bug Tracking	Problem Tracker
Regression	LSF
Memory Models	Denali

6.4   Transaction based verification environment:

Transaction based verification (TBV) allows simulation and debug of a design at the transaction level, in addition to signal/pin level. It enhances the verification productivity by raising the level of abstraction to the transaction level. Self checking and directed random testing can be easily performed instead of verifying using a pool of large test vectors or golden log files or reference models. In TBV environment you can create transaction verification modules (TVMs) that spawn the concurrent tasks. This makes it simpler to develop self checking cause-and-effect tests.

The TBV environment provides the following features:

- Generates self-checking and parameterizable testbenches at system or transaction level.
- Support for object oriented programming.
- Enhances Verilog and VHDL capabilities through the use of HVLs, which speeds test creation and maximizes reuse.
- Testbenches can be authored in Verilog HDL or other HVLs.
- Records, displays, and analyzes the response at system or transaction level in addition to signal/pin level.
- Debugging capability.
- Provides functional coverage of the design.

- Directed random testing to test the corner cases.

- Ability to reuse testbenches.

- Ability to partition and develop transaction verification modules (TVM) and tests by different verification team members.

6.4.1   Elements of TBV

The basic elements of the TBV environment are:

- **Design under test (DUT) or Design under verification(DUV)**: An RTL or gate-level description of the design that is to be verified.

- **Transaction**: A single transfer of data or control between a transactor and the design over an interface that begins and ends at a specific time.

- **Transaction verification module**: A collection of tasks that executes a particular kind of transaction. Each TVM can generate high-level transactions, for example, do_write, do_read, do_burst_write, do_burst_read, expect_write, send_data. The transactor connects to the DUT at a design interface. Most designs have multiple interfaces, hence, they have multiple transactors to drive stimulus and check results. TVM is also referred to as a transactor or bus functional model (BFM). TVMs can be authored in HDL or any other HVLs. TVMs created for a design can be reused for other designs involving the same or similar design blocks.

- **Test**: A program that generates sequences of tasks for each of the transactors in the system. The test program invokes transactor tasks indirectly and triggers the DUT. Tests can be authored in HDL or HVL.

As shown in Figure 6-2, the DUT is driven by the TVMs on its design interface. The design interface is basically a set of signals and buses. Usually a single finite state machine (FSM) controls the signals and buses in the design being verified. The sequence of states that is traversed during activity occurring on the interface consists of transactions. Examples of transactions include read, write, and packet transfers.

TVMs are driven by tests according to the system test requirements. The tests can control randomness of transaction sequence based on a previous transaction sequence, system rate, or coverage information.

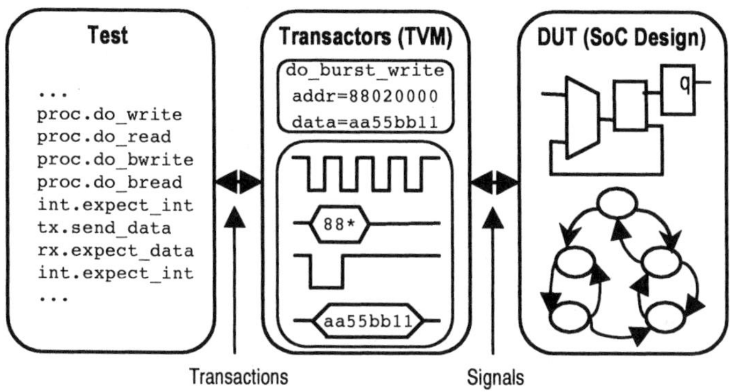

Figure 6-2. Transaction-based Verification Elements

6.5   Design verification

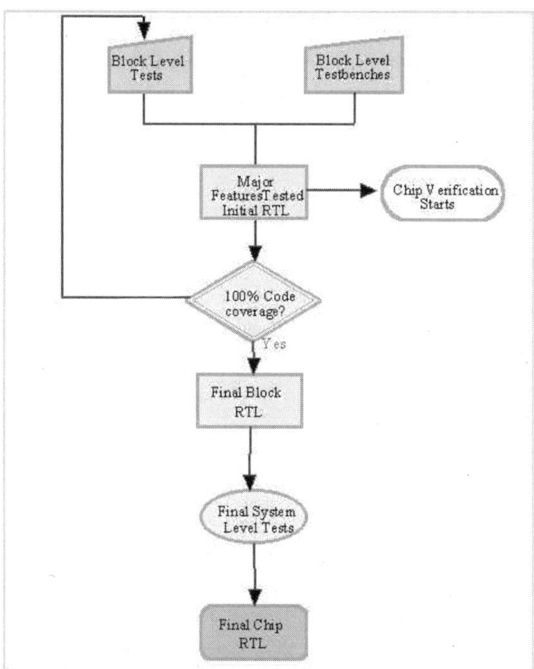

Figure 6-3. Block verification flow

In order to verify the complete design appropriate testing has to be done at both block level by design engineers and at system level by separate verification team. In a truly robust verification environment, block level verification is also done by verification engineer. There should be verification methodology clearly distinguishing the tasks for block level and top level verification. More focus for unit level tests and block features is at block level itself. After all the blocks are passing the block regression marked by 100% block code coverage, and the required limit for functional coverage is met, the RTL should be released to the system level verification.

As shown in the flow figure 6-3 at block level initial testing and code coverage for the RTL is done. When code coverage and functional coverage goals are met at block level, final block RTL is released to the system level testing which results in final verified chip level RTL.

6.5.1 System Level Verification

The system level testbench provides the verification environment to test the chip at top level. As in an example figure 6-4 there can be multiple interfaces supported by the chip for example - SPI4, PL3, PL2, UTOPIA 2/3, CSIX, NPSI and the CPU interface - driver TVMs will be required for each of these interfaces. On both sides of the chip are input and output interface drivers. The traffic generators drive traffic through the TVMs for all these interfaces. Test stimulus can choose one or multiple interfaces to drive traffic to.

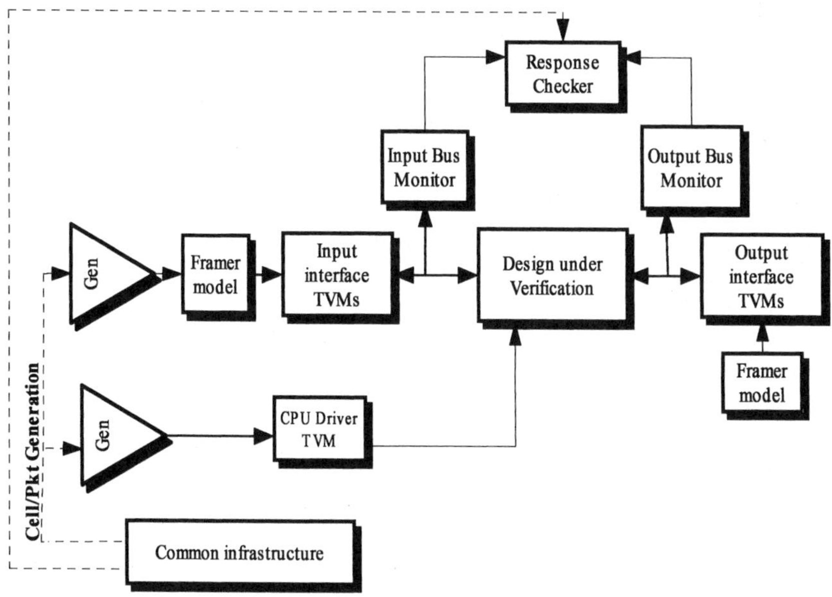

Figure 6-4. System level verification

Besides these driving interfaces, there is also CPU TVM to drive the cpu commands for chip configuration. At each interface there is attached a monitor besides the generator. Checker talks to input and output monitors of the chip/block, predict the expected behavior and compare it with actual behavior. Test could optionally also log the traffic activity and error messages into the corresponding log files. Diagram in above figure 6-4 gives overall picture of the verification structure at system level. Common infrastructure has common code for generation/checking and monitors. The diagram is applicable to any block level also. For block level Interface TVMs and monitors will specify the block interface protocol.

The DUV will have various internal bus monitors to aid system debugging. These are present on the interfaces of the blocks to track the state of the packet/cell as it goes through the system. These monitors act as passive monitors and record the activity into the SST2 database. Every block in the chip will have monitors at input and output interface. These monitors will provide following function:

- Track the cell/packet through the device at each interface.

- Log the information and pass it on to checkers.
- Flag any error against any interface conditions being checked on interface signals.
- Can flag EOP/SOP errors.
- Can flag Parity errors.
- CRC errors, bad packet.
- Report the rate of traffic for time window.

Basically, each monitor besides passing on the cell information to the checker will also pass on any error information to be acted upon by central messaging task. Any other checks or info that central messaging task needs from the block interfaces will be provided by monitors

Once the environment is setup to run one instance of system level verification, it can be duplicated for other instances and even for block level. For example if the interface change at the top level from serial to parallel bus, we just had to rewrite TVMs at the input/output, rest of the checkers, generators and tests remained the same.

6.5.1.1 Verification group deliverable for System Level verification

- System Verification Plan.
- Chip Level TVMs.
- SPI4/Pl3 Monitors.
- Block Bus Monitors.
- Packet/Cell Generators.
- Packet/Cell Checkers.
- Memory Violation Checks.
- Interface Models.
- Function Coverage.
- Regression Tests.
- System Tests.

6.5.1.2 Design group deliverable for system level testing

Design group has to contribute to system level tests and review of the system level test plan. They can identify some of the system level corner case tests that are not easy to create at the block level interfaces.

- System level tests.

6.5.2 Block Level Verification

The environment described above for system level verification is true for block verification also. Checkers for all the blocks further increase the level of confidence once designers have verified the block.

Block verification is mainly the responsibility of designer who owns the block. Individual designer is responsible for

- Block test plan.
- Test stimulus.
- RTL Lint checking.
- Assertion based checking.
- Synthesis check.
- gate level simulation.
- Static timing analysis.
- Code coverage.
- Model checking.

System level verification infrastructure should be developed in a way to reuse most of the components at the block level for block level testing. These are the infrastructure components required for block level testing which verification team can provide to block designers to aid in testing:

- Block interface Transaction verification module: This is the driver transaction protocol to drive the stimulus into the block interface. This will be required for all blocks that have to be verified individually. These interfaces can be also used at system level testing if some of the block RTL is not ready. For example if external interface to the chip is not ready but most of the internal blocks are ready than stimulus can be applied directly to one of the internal interface.

- Block Bus monitors: Bus monitor are developed for system level testing to track the data progressing through the block interfaces. The same monitors can be used at block level testing also.

- Packet/Cell Generators: The same generator infrastructure at top level can be used at block level. There has to be an option to attach generation to certain interface through driver TVM.

- Memory violation checks: Memory violations checks can be done faster from blocks that are attached to the external memory.

- Block response Checkers: Checkers developed at top level can be used for block level also.

- Coverage/Run scripts: Block level verification also needs the code coverage and regression running scripts. Most of this can be reused from the top level verification.

6.5.3   Block TVMs at system level: Interface TVMs:

Here are the main features of block TVMS:

- Traffic is generated at the interfaces via interface driver TVMs. For example in figure 6-4 input interface TVMs are driver TVMs for SPI4, PL2, PL3, UTOPIA2, UTOPIA3, CSIX, NPSI or all combinations of these interfaces as supported by the chip.

- Traffic can be generated at any block, incase chip interface changes, only TVM has to be re-written

- If previous block is not available, generation & checking can hook up to the available blocks at system level

As shown in the figure 6-5 below there are bus Monitors at each Block interface at System Level. One of the advantage of having all the block TVMs, Monitors and checkers is that at system level we can hook up these monitors or checkers/TVMs to closely look at the block interface for debugging. Also, if block RTL is not ready, verification at system level can continue by replacing it with the TVMs for that block. Following figure 6-5 represent that if input interface is not available, then generation (driver) TVM for Lut engine can be used to drive the stimulus.

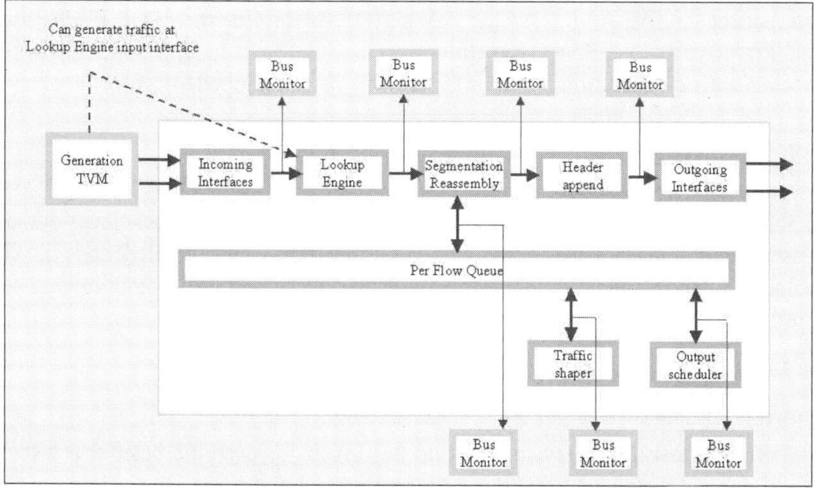

Figure 6-5. Bus monitors at block interfaces

6.6   Verification infrastructure

Verification infrastructure consists of following.

- Interface TVMs: Driver transaction verification modules for all the blocks and the top level chip interface.

- Traffic Generation for cells or packets: Traffic generation code for generating various networking traffic types.

- Monitors: Monitors at all the interfaces.

- Checkers: Checkers for all the blocks in the chip.

- Messaging task: To pass on any messages from monitors or checkers to common message responder that can take an action on these.

- Memory Models: Dynamic memory models for memories used in the chip to be used for functional simulation.

- Common data structures: For various generation types, common monitor tasks, common checker code and generator code.

The appendix provided at the end of the chapters discusses some common data structure developed using testbuilder (initial version of SCV) environment to be able to verify any networking Design. It implements networking infrastructure of cell, packet – ATM, AAL1, AAL2, AAL5, L2, MPLS, Ethernet, POS(PPP/IP) format, Frame relay over Sonet data structure. Common SAR functions, rate monitors, checker, traffic generation, monitor tasks were also developed.

6.7   Interface TVMs

In this section details of interface driver TVMs for chip or block level are discussed. Examples are given for standard interfaces. The detailed code is also included in the Appendix. An interface driver TVM should have following features:

- Model the protocol for the driving interface accurately.

- Option for starting or stopping the generation at any time.

- Different task for every interface to be driven for example besides a task for driving data and control path signal, another task for driving the training pattern in SPI4 and for driving the back pressure etc.

- For all the signals that are don't care, must drive the random data to make sure that there is nothing wrong in implementation of the RTL with respect to these don't care signals.

- For the behavior that is optional make sure the driver TVM supports all the options and different tests choose randomly between these options. For example, it is optional in generating traffic for PL3 PHY interface to send packets without port indication in between if all packets are on the same port. Port indication can be just set once and all packets on the port can be sent with or without port indication. There can be bug in the RTL which is not modeled keeping this option in mind.

- Should be able to generate any error conditions in the interface. Should be able to take an option of ratio of good traffic vs. bad traffic on the interface

- Timing must be specified as per standard interface spec in the driver TVM, incase the driver interface TVMs are to be used for gate simulation

An interface driver TVM once designed as above is reusable and can be used in any design, independent of the current project. Following sections show examples of some standard driver TVMs that can be used in any design that is using the standard interfaces irrespective of the verification environment being used, since testbuilder is open source code and can run with most of the known simulators.

6.7.1 PL3

This section discusses the implementation of driver interface TVM for standard PL3 PHY-LINK interface as defined by optical inter networking forum OIF-SPI3-01.0 spec titled "System packet interface Level 3 (SPI-3):OC48 System Interface for Physical and Link Layer Devices". This spec specifies the PHY-LINK interface. It describes a packet interface of OC48 aggregate bandwidth applications. The interface supports bidirectional data transfer between two devices, through two different set of signals that operate independently from the other.

The SPI-3 interface defines the interface between SONET/SDH Physical layer devices and Link Layer devices, which can be used to implement several packet-based protocols like HDLC and PPP.

Following figure 6-6 is an example of how single multi-port PHY device can be interfaced to Link layer device. In the example, the Link Layer device is connected to a single package four channel PHY layer device using the 32-bit interface. For more details please refer to above spec in www.oiforum.com.

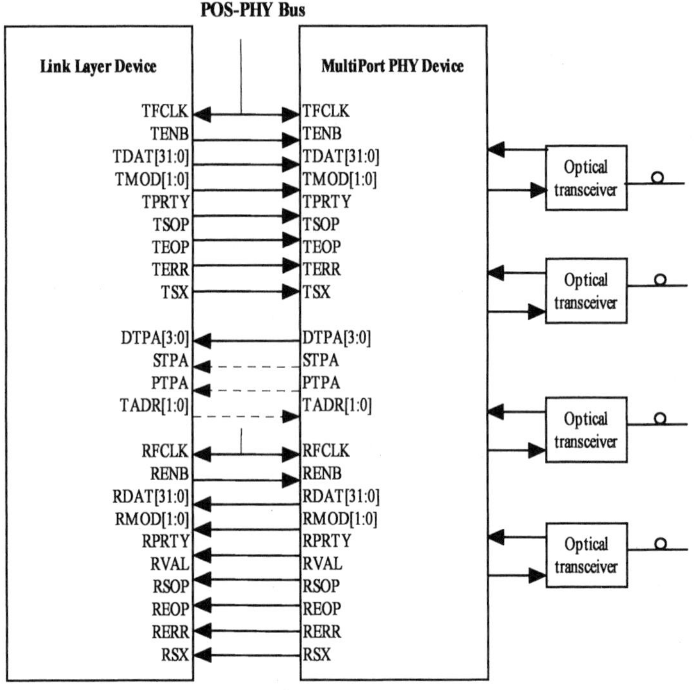

Figure 6-6. PHY to LINK Interface

6.7.1.1 PL3-Tx Description:

The SPI-3 transmit interface is controlled by the Link Layer device using the TENB signal. All signals must be updated and sampled using the rising edge of the transmit FIFO clock, TFCLK. Figure 6-7 is an example of a multi-port PHY device with two channels. The PHY layer device indicates that a FIFO is not full by asserting the appropriate transmit packet available signal DTPA. DTPA remains asserted until the transmit FIFO is almost full. Almost full implies that the PHY layer device can accept at most a predefined number of writes after the current write.

If DTPA is asserted and the Link Layer device is ready to write a word, it should assert TSX, de-assert TENB and present the port address on the TDAT bus if required. Subsequent data transfers with TENB low are treated as packet data which is written to the selected FIFO. At any time, if the Link Layer device does not have data to write, it can de-assert TENB. The TSOP and TEOP signals must be appropriately marked at the start and end of packets on the TDAT bus. When DTPA transitions low and it has been sampled, the Link Layer device can write no more than a predefined number of bytes to the selected FIFO. In this example, the pre-defined value is two double-words or eight bytes

Figure 6-7. Transmit Logical Timing

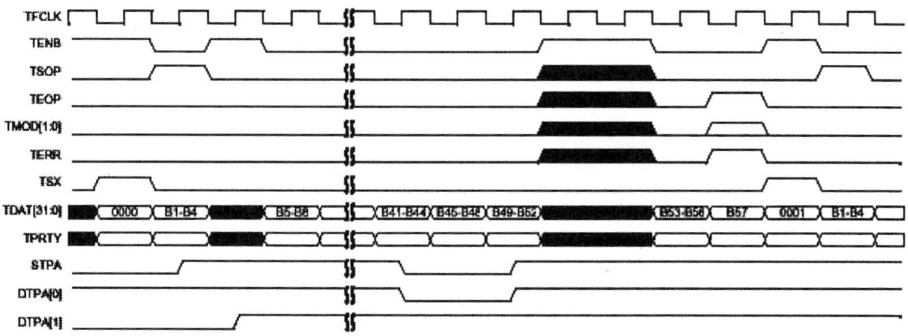

6.7.1.2 PL3-Tx Implementation

This section discuses implementation of driver TVM for above protocol in test-builder using C++ based verification methodology. As described in the figure 6-8 below, this consists of two concurrent processes (two tasks)

- Status Task: Following is the functionality of this task:

1. Drives output port values, according to the port calendar, polling for the status for those ports.

2. Port status is collected 2 clocks later.

3. Store status and use it as reference while sending data.

4. To prevent status collection task from running forever, a status collection time interval is defined through a boolean variable set and kept while generation continues and then reset.

- Send Data Task: This process preforms following functionality:

1. Check the status - if it allows start sending data, else wait for status.

2. Send data - There are 2 modes of sending data - Blocking and Non-Blocking.
    Blocking: If one of the ports is blocked, traffic is blocked to the interface
    If the status polled and coming back on ptpa is blocking a corresponding port -
    than wait for status change
    Non-blocking: If one of the port is blocked, then traffic continues on rest of the
    ports.
    In this case all transfers go into a FIFO and the only ones that have appropriate
    status are performed. The transfers for currently blocked ports are checked at
    every TVM spawn/run.
    After the last spawn/run only a change in the status (ptpa) causes a check for
    every transfer in the FIFO and triggers the data being sent out.

Flow:

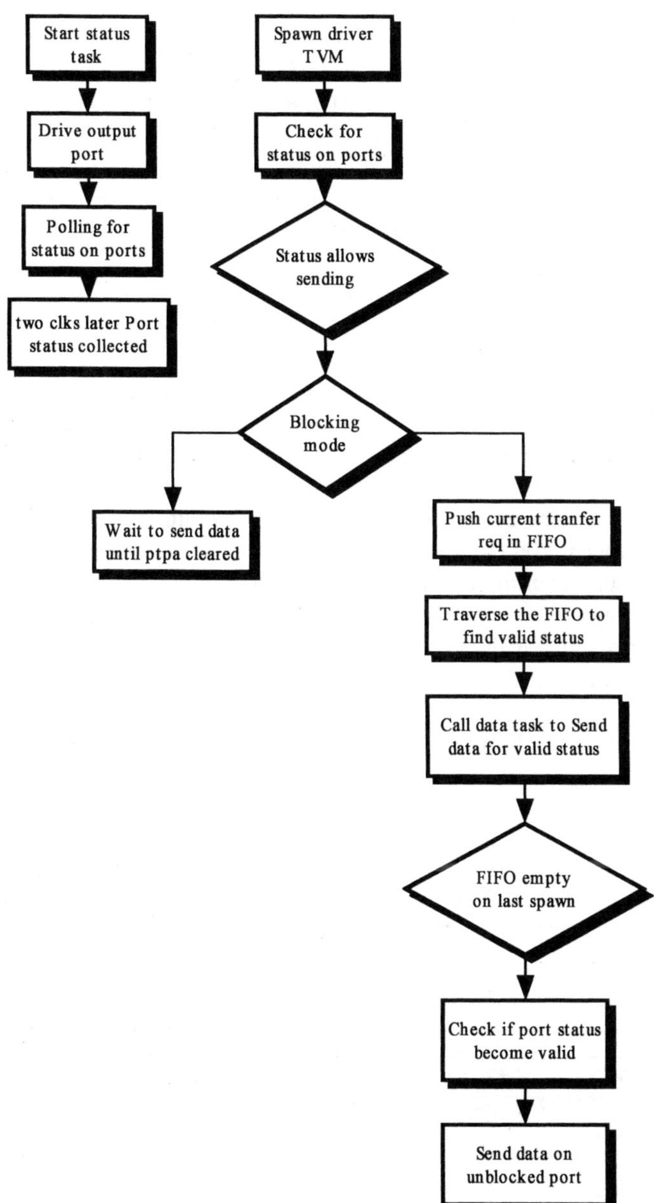

Figure 6-8. PL3-Tx flow diagram

6.7.1.3 Pl3-Rx description

The SPI-3 Receive Interface is controlled by the Link Layer device using the RENB signal. All signals must be updated and sampled using the rising edge of the receive FIFO clock. The RDAT bus, RPRTY, RMOD, RSOP, REOP and RERR signals are valid in cycles for which RVAL is high and RENB was low in the previous cycle. When transferring data, RVAL is asserted and remains high until the internal FIFO of the PHY layer device is empty or an end of packet is transferred. The RSX signal is valid in the cycle for which RVAL is low and RENB was low in the previous cycle.

Following figure 6-9 is an example of a multi-port PHY device with at least two channels. The PHY informs the Link Layer device of the port address of the selected FIFO by asserting RSX with the port address on the RDAT bus. The Link Layer may pause the Receive Interface at any time by de-asserting the RENB signal. When the selected FIFO is empty, RVAL is de-asserted. In this example, the RVAL is re-asserted, without changing the selected FIFO, transferring the last section of the packet. The end of the packet is indicated with the REOP signal. Thus, the next subsequent FIFO transfer for this port would be the start of the next packet. If an error occurred during the reception of the packet, the RERR would be asserted with REOP. Since another port's FIFO has sufficient data to initiate a bus transfer, RSX is again asserted with the port address. In this case, an intermediate section of the packet is being transferred.

Figure 6-9. Receive Logical Timing

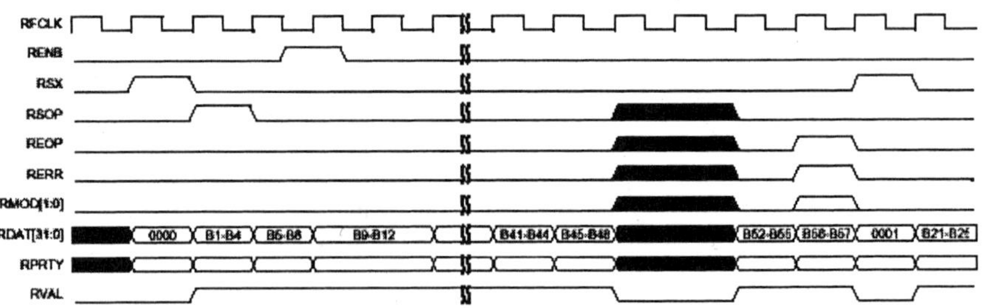

6.7.1.4 Pl3-Rx Implementation

This section discusses the implementation of driver TVM for above protocol. As seen in flow diagram in figure 6-10 there is one task for this TVM which is spawned in the beginning. It waits for enable to be asserted from Transmit side and drive Sx and Port signals. Next it drives data and control signals (RMOD, RSOP, REOP, RERR, RVAL). This is done after checking with each clock if enable is asserted. When it is last byte transfer, Valid is removed after last valid byte is sent.

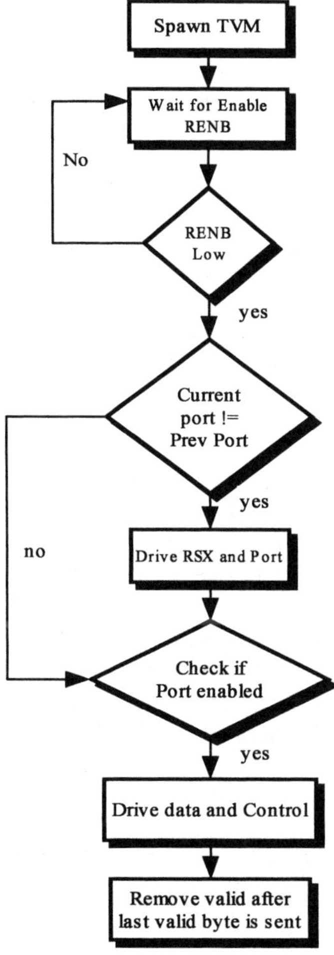

Figure 6-10. PL3-Rx flow diagram

6.7.2 Other interfaces

Most of the other standard interfaces mentioned in the beginning of this section, are similar to the above interfaces PL3 and SPI4 that have been discussed in detail. There are only few differences between PL2 and PL3 and UTOPIA 2 and UTOPIA 3. In case design supports all the interfaces, it is a good idea to encapsulate the protocol in one TVM which has most of the code similar with options for choosing among different interfaces for the design. There can be common TVM for all of these interfaces and any of the interfaces can be triggered on while generating the traffic.

- PL2
  PL2 interface is similar to the PL3 interface except that cell transfer can be interrupted in the middle for PL3 whereas in PL2 whole cell has to be sent out.

- Utopia
  Utopia interface is for ATM. Similar to PL3 interface except that PL3 is for packet. Difference between UTOPIA 2 and UTOPIA 3 is that in UTOPIA 3 cell transfer can be interrupted in between whereas in UTOPIA 2 whole cell has to go out.

Detailed example for SPI4 interface is also provided in the appendix.

6.8 Traffic Generation

Every chip verification requires a stimulus generation code. For networking design, the packet and cell generation are the primary stimulus to the design under verification. They are implemented in bus independent manner so that traffic can be sent on any bus interface. This section focuses on the methodology and implementation for the traffic generator for driving chip stimulus. It is designed to emulate "real life" network traffic and can be extended to include any Packet generation for POS(PPP/ IP), Ethernet packet format, Frame Relay over SONET etc. The general methodology and implementation are discussed in following sections.

6.8.1 Traffic generator methodology

Traffic generator should have following features:

- Stimulus generation for networking traffic commonly required for network chip verification.
- Can be extended to include any new types of traffic generation.

- Defines ATM cell data structures, AAL5, Ethernet packet, switch packet data structures.
- Defines various functions to be able extract different fields to generate directed tests.
- By default if field is not specified is the test, it is generated randomly within the constraints.
- Can vary rate of incoming traffic based on leaky bucket parameters or any other mechanism.
- Can specify options for which block or interface to generate traffic on.
- Which ports/flows generation is targeted at.
- Other parameters related to the flows or cells that are generated.
- Number of cells/packets to be send in.
- How many simulation cycles to run.
- Specify when to start or stop the generation.
- Empty cell insertions.
- Error introduction and insertions.

For selecting a particular format of cell to be sent on the port or flow at certain rate, a traffic scheduler is required that selects and schedules when to send a cell for a flow on a port. Besides traffic scheduler a data generator is required which composes and provides actual format of the cell to the traffic scheduler.

traffic scheduler: This is mainly the control path of traffic generator. It controls the rate of flow, generates the cells and calls appropriate interface in the chip to send the cell to. It selects a port and the flow that is to be sent out on that port. Once flow is selected, the data generation code is used to get a data fragment for that flow. Once data is available it will be send out to the block interface it is targeted for. This code will also keep track of the statistics for the flow and will also require the knowledge of port being in the middle of packet or not.

Data Generator: After traffic generation code has selected a flow, it will call data generator with the flow and port to send traffic on. Data generator will return data in form of a cell or a packet fragment to be sent out on the generating interface. To work properly, data generator also needs to know whether flow is in the middle of a packet. It decides how to rotate the port and flow for the next cell.

Stimulus file selects what type or category of traffic to be sent. For example, to be able to send AAL5 traffic or ATM or Ethernet packet. Data generator will figure out what headers or pads to attach to compose a cell conforming to the traffic type required. For data generation code, everything it needs to know about the flow is available from the dataGram (data structure carrying data and side band control information).

6.8.2 Overall Flow

Traffic scheduler determines whether to send the traffic or not based on credits that are calculated, and if complete packet has been sent or not. It also decides about when to stop the generation. The data generator deals with the data manipulation, sending different output types, negative traffic etc. There can be one class or category of class for each kind of traffic type to be generated. Any new generation type can be added to this code.

At the beginning of test, traffic scheduler is populated with the information regarding number of flows, Ports, Qos, output traffic type to be generated, interface TVM it is hooked up to, number of cells or packets to be generated, traffic speed and burstiness. It is spawned once in the beginning of the test. At the start of test, it populates the credits for port, qos and flow. After credit is populated, it selects a port, qos within that port and flow within that qos. If it's not the first cell of the packet it continues with the previous flow. Incase data is not available to be send, it calls fillQueue which looks for datacompP for this flow. Based on the output type to be generated it creates a list of cells. If data is available to be sent, it calls sendCell with data from flow's queue. If there is not enough credit to send the cell out, it sends an empty invalid cell. After the cell is sent, credit is recalculated based on the cell that has been sent. Generation is finished if all the required number of packets or cells have been sent or simulation time has finished.

Flow for generation is as shown in figure 6-11.

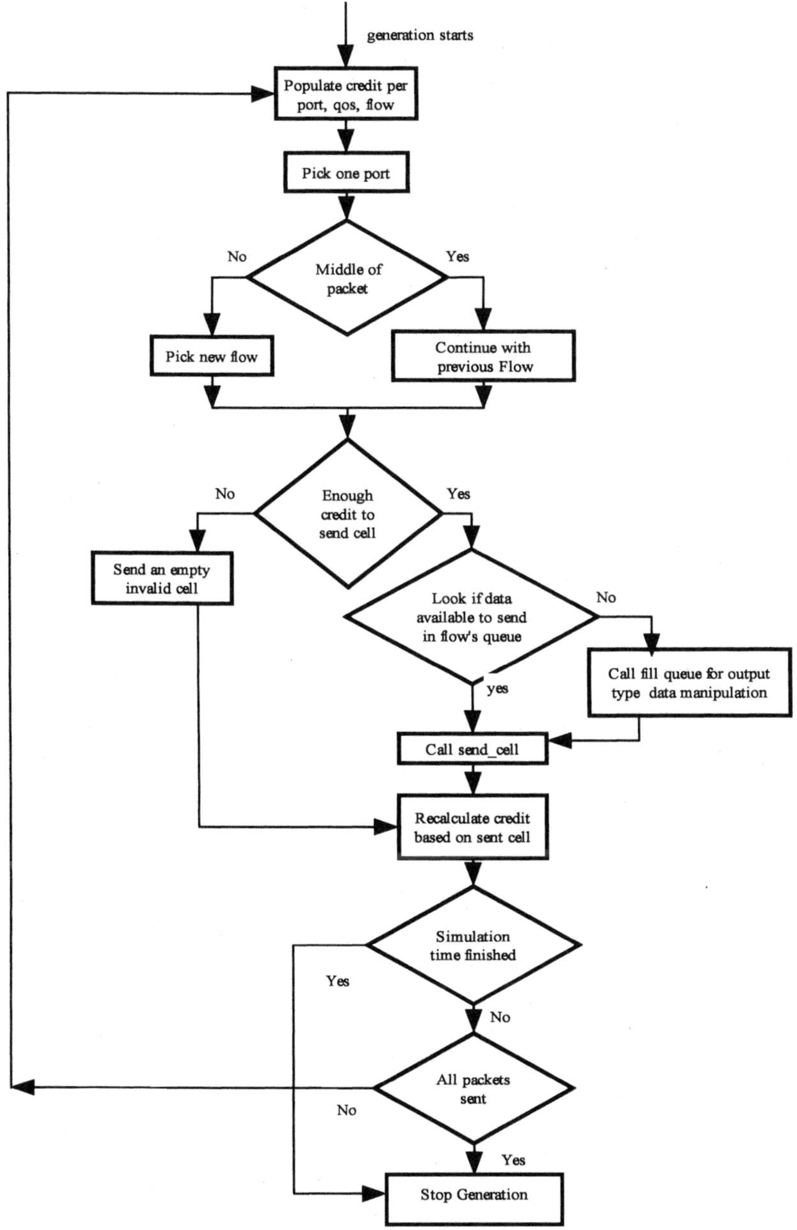

Figure 6-11. Traffic generation flow

6.8.3 Implementation flow

As one of the examples of above methodology implementation, flow diagram is shown in figure 6-12, the traffic generation code is divided into:

trafficScheduler.cc: Selects the flow and port to send cell on and sends the cell it gets form dataProvider to the chip interface.
leakybucket.cc: This keeps track of credits for the trafficScheduler to schedule next flow.
dataGenerator.cc: When trafficScheduler selects flow and port it calls fillsQueue from dataGenerator. dataGenerator calls getListofCells from the block data generator.
blockDataGen.cc: getListofCells is overridden by each specific interface dataGenerator. This decides to attach headers/pads, composes the actual traffic type with right format and pushes it to the Smart queue.

Traffic scheduler selects the flows, port to send the traffic. It calls the fillQueue from dataGenerator code which gets the actual cell from the block dataGenerator. Based on the traffic type to be send, block dataGenerator composes a cell and pushes it to the smartQueue. TrafficScheduler pops this cell from the smartQueue and put it to the mailBox (a queue with Semaphores) of attached task by calling pushIntoInterfacequeue.
pushIntoInterfaceQueue pushes the cell to the mailBox for the interface driver task. This is waiting for any cell to be pushed on to the queue and transform it according to the interface protocol signals.

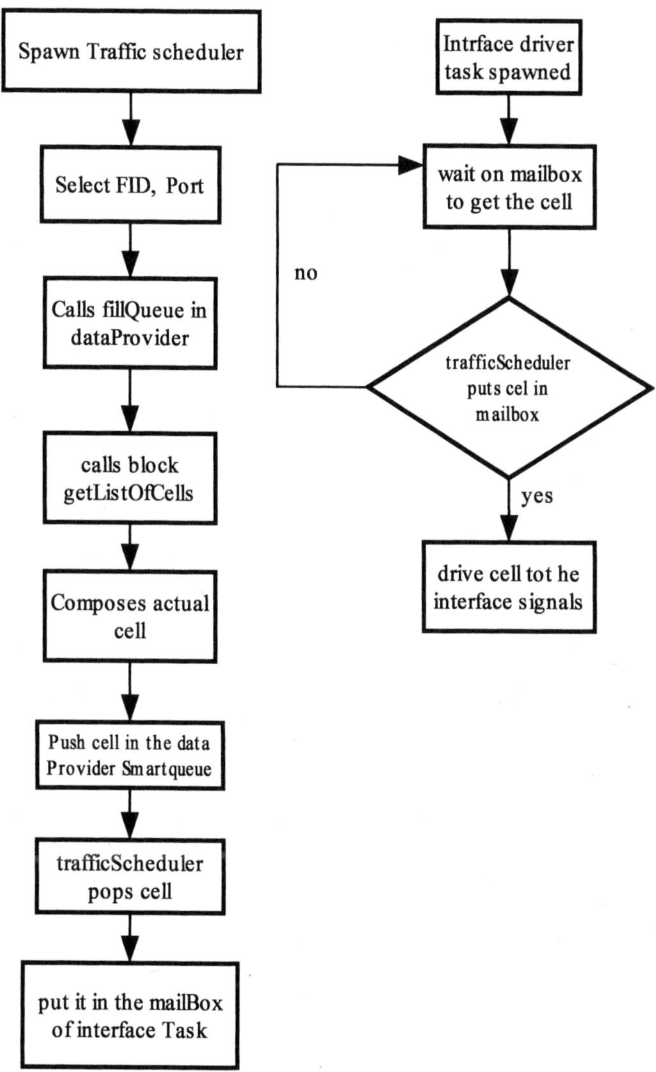

Figure 6-12. Generation implementation flow

6.8.4 Interface Driver hookup to traffic Generation

InterfaceDriver: interfaceDriver task is there to hookup between generator and drivers. It is the basedriver class that has interfaceDriverQueue (tbvMailBox) where the traffic Scheduler task pushes the dataGram that is to be driven. It also records transactions to fiber, starts and stop threads.

blockInterfaceDriver: This defines the TVM and the task to connect and drive the HDL signals and a task. The blockInterfaceDriverTask inherits from the interfaceDriverTask defined above. This task is spawned once in the beginning and it keeps waiting on the mail box until it is stopped. When mailBox sees cells to be driven from the trafficScheduler, this tasks picks the cell up and transforms to the interface protocol to send in the stimulus.

6.8.5   To setup test stimulus for generation:

```
//instantiate the block data Provider
  blockDataGenT * blockDataProvider = new blockDataGenT("block data pro-
vider");
// setup number of packets to be sent
  blockDataProvider->setTotalPackets(10);
//connect the block data provider to the trafficScheduler task
  trafficScheulerTvm.trafficSchedulerTask.connectDataProvider(blockDataPro-
vider);
// add the interface driver task to the trafficScheduler
 trafficScheulerTvm.trafficSchedulerTask.addTask(& blockDriverTvm.blockDriv-
erTask);
//Define the side band HDL signals (like traffic type, flow IF etc.) and set the traffic
in block data provider
 blockSidebandControlT * blockSidebandControl = new blockSidebandControlT();
blockSidebandControl->block_type = 0;
blockSidebandControl->block_qid = 10;
blockDataProvider->setupTraffic(blockSidebandControl);

//Spawn the trafficScheduler task and the block interface driver task
 trafficScheulerTvm.trafficSchedulerTask.spawn();
 blockDriverTvm.blockDriverTask.spawn();
```

6.9   Writing Stimulus File

Key to writing efficient testbench for any regression is to be able to write complex test scenarios in easy and simple format. For complex chips, thousands of tests have to be written that are regressed over months to prove the functionality of the chip. Stimulus file format should be as easy as it can be. Underlying verification infrastructure of generators, checkers monitors, chip configuration can be very complex and have lot of features but unless the tests can be written in simple form by anyone who is verifying the chip - the whole purpose is lost. If high level language like c/c++ is being used, an intelligent front end can be written for writing stimulus files to be able to describe complex random cases in concise format. Some engineers also use scripting language like TCL or perl to be able to describe the test stimulus. As the infrastructure grows and more and more features are added to the chip and verification, describing new test scenarios in the stimulus file should be supported.

Based on the chip features the test stimulus API can get very complex. For example in the example reference design there is lot of setup and constrained randomization required. For a chip where s/w is more integrated with the design stimulus generation can just mean reading from the image or data file and generating packets from it.

Here are some main features that stimulus files should have:

- Random generation: By default if parameter is not specified, it should be randomized within the defined constraints. For example, if stimulus file is just one line which say generate one million packets, it should pick up all the rest of parameters randomly like which interface to generate, what size of packet, what chip configuration etc. For all the values that are randomized, the information should be logged in some form that state the actual value assigned. There should be seed information logged to be able to rerun the test with same parameters.

- Support events: It should be easier to generate any event in the middle of the simulation run. For example, while simulation is running and after seeing generation of hundred packets, it is required to block the traffic on one of the modules to generate backpressure for debugging purpose. Event to be generated while simulation run, can be a simple wait statement or chip configuration change by programming registers or some complex function describing the behavior.

6.9.1   Example stimulus file front end

This section discusses the stimulus file format for verifying the reference design that used testbuilder as verification tool. This should be reference for anyone who needs to write the tests for the RTL verification.

Overall Structure:

Stimulus file code is mainly calling publicly available API function calls. Taking an example from the same reference design and Testbuilder, this code consists of two main categories

- one which operates on test object lets say testCaseT
- second which operate on description of flow ID on which particular traffic comes in lets say flowDescriptorT or related objects like different RTL fields to be configured for example rtlFiedlsT.

This code operates on the flows and RTL, test constraints. There is one testcaseT object and one or more flowDescriptorT object. A stim file itself is of the form of a testBuilder tbvMain function.

Following is example of minimal structure of a stim file with no traffic or function calls. It will randomize everything:

```
// to define the seed for the test in case needed to be rerun
#define INIT_SEED5555
//macro file that contains all required includes
#include "stimLib.h"


tbvMain ()
{
//this line is required to instantiate a test
testcaseT   newTestcaseObject;
//should match file name
newTestcaseObject.testNameStr = "new_sample_test";
newTestcaseObject.testNameDescription = "To Clarify stim basic structure";

//other useful test stimulus code

newTestCaseObject.Run_Test(); //To run the test
tbvExit();  //To stop the verilog simulator
}
```

The testcaseT and flowDescriptorT object classes have a number of modules of functions and members which constitute the functionality of the stimulus file environement. The API functions are generally scattered over testcaseT itself and its members. Following are the four members of testcaseT:

6.9.1.1 API Function Overview for testcaseT:

Other than the two strings for testname and test description, this level contains only the API functions which run the test and instantiate flowDescriptorT or rtlFieldsT objects for example newTestCaseObject.newflow();

- testcaseT.rtl: This member of testcaseT is of the class rtlLibT(RTL library) and contains all API functions which are used to set or get parameters in the hardware not specific to a flow or group of flows. In general the functions are of the naming form setXXX Parameter, where XXX is the rtl module abbreviation in which Parameter is stored. Note that not all public members of this class (rtlLibT) are considered part of the stimulus file API. In general, this class provides the most API function calls. for example:

newTestcaseObject.rtl.setSCHMode(some_unsigned_int);//set mode for scheduling

- tescaseT.traf: This member is used to hold all settings related to the generation of traffic patterns through the device under test which are not specific to a flow. This class (trafDescT) has other public functions which are not considered part of the stimulus file API. The API functions that are present generally set variables that can be set on individual flows. In general, the variable as set on the flow takes precedence only if explicitly set in the stim file. for example:

newTestcaseObject.traf.setPacketSizeMax(some_unsigned_int);//set   maximum packet size

- testcaseT.bench: This member has some public functions which are part of the stimulus file API as well as some members variables which can be set by user directly. The primary purpose of this module is to control behavior of the underlying common testbench engine by providing a way to control the simulated environment surrounding the device under test. Any data or function call which affects anything other than the configuration of the device under test should go here. A lot of the functions in this module have non-trivial requirements, so API wrapper can be provided using the testMethodsT library. for example:

newTestcaseObject.bench.setSPIGen();//set generation on SPI4 interface

- testcaseT.testLib: This class (testMethodsT) is for any functional API calls which do not fit neatly into the other three sub sections of testcaseT. These functions will mostly be testbench features which need access to multiple "domains" of data in order to function properly (such as back pressure). These functions generally assemble other public API functions into a higher level behavior. The only functions, member of this class may operate on are other public, API functions. In other words, any function in this library could be done using the discrete steps in the stimulus file, but it may be easier to code a general procedure for those steps in this object. for example:

newTestcaseObject.testLib.setupBackPressure(constLibT::SPI4, 10000);//setup backpressure for 10000 ns on SPI4 interface

6.9.1.2   API Function Overview for flowDescriptorT and rtlFieldsT:

rtlFieldsT objects and their derived class flowDescriptorT objects provide an API for configuring shared resources through the use of the underlying configuration solver. ALL flowDescriptorT or rtlFieldsT objects must be declared explicitly using the functions in testcaseT, since these objects have extremely non-trivial constructors. The testcaseT module provides a mechanism for returning references to flowDescriptorT or affiliated objects which can then be operated on as references.

In general, flowDescriptorT objects have only local scope (e.g. setting a keepOut on a field of one flow object has no effect on another flow object) whereas rtlFieldsT objects come in two scopes local and global. Global scope rtlFieldsT objects are extremely powerful, so extreme care should be taken when using them. Most of the useful functionality of these classes is provided through the manipulation of data within these classes rather than through API calls on them. For example:

someFlowObjectReferenceDeclaredElsewhere.lutPort = 5;

Both the flowDescriptorT and rtlFieldsT data members should be assumed to support the same API as the tbvSmartUnsignedT class. Note that the following are considered equivalent with respect to the potential state set.

    myRtlDescTReference.lutPort = 5;
    myRtlDescTReference.lutPort.keepOnly(5);

The field operations for these objects are all processed through the Randomization engine code that will attempt to reconcile all of the settings of all of the objects in both local and global scope and come up with a testcaseT RTL configuration which satisfies all of the settings. If no configuration can be found in a reasonable number of tries, the engine will assert (false) and stop the test.

Any operations on rtlFieldsT objects with global scope will be applied to all flows and constraint objects across the entire test. Global scope rtlFields object are mostly provided as a computational efficient way of masking unwanted configuration options from use. It is anticipated most of the field operations on global scope rtlFieldsT objects will be on non-contended resources trafficType(type of traffic like ATM, AAL5, packet etc.), inputInterface(Inputinterface to the chip like SPI4, PL3, UTOPIA etc.) and outputInterface(outputI interface to the chip like SPI4, PL3, UTOPIA etc.)

Note also that flowDescriptorT objects have a special property with respect to actual flows as programmed into the RTL. A single flowDescriptorT object may represent one or more actual flows in the final testcase. The size of these groups is set at construction time.

6.9.1.3 Complete API:

All functions which do not explicitly state their return type can be assumed to have return type void. The stimulus file author need not be at all familiar with the inner workings of the code, just the API calls for the functionality that are of interest for the test:

rtlFieldsT and flowDescriptorT API:

rtlFieldsT data fields which can be manipulated in a stimulus file:

```
inputInterface { SPI4, POS3RX, POS3TX, UTOPIA, CSIX}
spiiPort {0..63} // port number for input interface
lutPort {0..63} //port number for look up block interface
trafficType {0,1,2,3,4,5,6,8,9,10,11,12,13,14} //incoming traffic type
rasPort {0..63} // reassembly block port number
spioPort {0..63} // output interface port number
outputInterface {SPI4, POS3RX, POS3TX, UTOPIS, CSIX} // output interface
virtualPort {0..255} // depending on mode of scheduling
```

Note that the encoding for SPI4, POS3RX and POS3TX are stored as static, public data in constLibT.h, so the normal use of these symbols is of the form: constLibT::POS3RX. Also, setting a given resource on a particular flow or rtlFieldsT object actually affects a constraint in the underlying configuration solver. This means that any other flows which happen to solve into the same configuration space end up in some sense sharing that resources properties. For example setting a lutPort to inputInterface relationship with respect to one rtlFieldsT object means that another flowDescriptorT which uses the same lutPort will also solve to the same inputInterface, even if this is not explicitly stated.

API of rtlFieldsT objects of global or local scope:

1. operator=(rtlFieldsT &): This function copies one rtlFieldsT object onto another, for instance if there is a complicated set of input side constraints and different output constraints, it may be easier to set the input side constraints on one rtlFieldsT object and then make two copies, then set both copies output sides separately.

2. operator<<(): This prints out the contents of an rtlFieldsT object. Use with caution, as most underlying fields won't be set at this stage in the stim file, so the info produced by a print in the stim file isn't guaranteed to be useful or final.

API of individual fields (spii, lutPort, etc...) :

1. keepOnly(unsigned int state): This function holds the potential value of the field to just the value of state. This function will blow an assertion if the value of state is normally illegal for this type of field. This function will also assign the current value of the field to the value of state. It is actually quite similar to operator=(unsigned int state).

2. keepOnly(unsigned int lower, unsigned int upper): This function holds the potential state of this field to whole numbers between lower and upper, inclusive. This function will still work if lower == upper.

3. keepOnly(set<unsigned int> & stateSet): This function will limit the field to use only states that are contained within the set of stateSet. This function does not currently do error checking for normally illegal states, it simply copies the states into the internal set of states.

4. keepOut(unsigned int state): This function will remove the single value state from the available states for the field. If state is not a currently valid value for this field, this call will be ignored.

5. keepOut(unsigned int lower, unsinged int upper): This function removes all integers between lower and upper inclusive from the set of available states for this field. Any values which are already missing from the state set are ignored.

6. keepAdd(unsigned int state): This function inserts the value state into the set of available states for this field. It does not check if this state is ordinarily valid for this type of field. This behavior is useful for testing conditions which are pathological or illegal. The various flavors of keepAdd are generally useful for constructing highly customized sets of values for certain fields or for testing normally illegal values.

7. keepAdd(unsigned int lower, unsigned int upper): This function inserts the values from lower to upper inclusive into the set of available states for this field. It

does not check if this state is ordinarily valid for this type of field. This behavior is useful for testing conditions which are pathological.

8. randomize(): This function selects a random state from among the currently valid states available to that field and assigns it to this field. Note that this function is useful for assigning multiple fields across objects to a randomly assigned, but equivalent value. Note that this function does not affect the available states. There is no public randomize() function for an entire flowDescriptorT object as the randomization process is non-trivial. Since this call does not affect the state of the field to which it is applied, it is also handy to follow up this function with a freeze() call, else the later randomization process will overwrite the first randomized state value.

9. freeze(): This function will reduce a field to only keep to it's current state. Assuming the field has a current value (generally given by randomize()), then the following are equivalent.
someRtlFieldTObject.someField.freeze();
and
someRtlFieldTObject.someField.keepOnly(someRtlFieldTObject.some-Field.getUnsignedValue());

10. operator=(unsigned int value): This function is equivalent to keepOnly(value);

11. setState(unsigned int state): This function sets the current value of the field to the value state. It does check to make sure that state is currently valid for this field.

12. operator=(const objectOfSameType & rhs): This is equivalent to a copy constructor for the fields involved. In general only fields of the same Type may be assigned (i.e. lutPort needs to be assigned to other lutPorts, etc.).

13. unsigned int getUnsignedValue( ): This function returns the current value of the field. Note that this function's behavior is undefined if the field has not had a value assigned to it.

flowDescriptorT Operations

The flowDescriptorT is a child class of rtlFieldsT and therefore supports all of the field based Operations of the rtlFieldsT base class. flowDescriptorT itself also has classes derived from it, but in general, these support all the operations below as far as the stim file is concerned. A flowDescriptorT object in the stim file may represent a large group of flows once the data structure is processed by the test code. These flows will all share the same basic settings, but any setting not explicitly set or constrained will still be random across its legal range. Flows which are declared with no group size default to having a group size of one.

Fields available in the flowDescriptorT in addition to base class fields:

```
rateID e {0..255}// Rate id for shaper block to shape traffic
schOn e { 0, 1 }// to switch on or off scheduling
shpOn e { 0, 1 }// to switch on or off shaping
packetSize e { 1..any size } // size of packet
```

Additional API functions for this class:

1. Following functions provide the interface to the flow's FID(Flow ID number)Tag. The valid range of FIDs is configurable in the hardware, therefore, this range may change. Since the flow's FID Tag also affects it's valid shaper status, this function is aware of the valid shaped flow ranges as well. If the user sets the flow to a region outside the shaper range, the setFID() function will also turn off the shaper for this flow by default.
   unsigned int getFID();
   void setFID(unsigned int fid);

2. Following functions control the scheduler behavior for this flow or group of flows. Alternatively, the schOn field may be operated on with the standard keepOut, keepOnly functions for field data. Note that for the function setSchedulerEnable, the parameter enable is a flag which may be true or false.
   enableScheduler();
   disableScheduler();
   setSchedulerEnable(bool enable);
   bool getSchedulerEnable();

3. Following functions get and set the flow's QoS within it's virtualPort.
   setSchedulerQOS(unsigned int value);
   unsigned int getSchedulerQOS();

4. Following functions set the status of the flow with respect to the shaping, policing block. Note that these functions have side effects which restrict the range of the flowID tags available to this flow.
   enableShaper();
   disableShaper();
   setShaperEnable(bool enable);
   bool getShaperEnable();

5. Following functions are equivalent to operating directly on the rateID field of the flow.
   setRateID(unsigned int _rateID);
   unsigned int getRateID();

6. Following function sets the number of cells which will be generated on each flow of this group. If this function is called on a flowDescriptorT object, it will

use the value set by this function instead of the default set by the trafDescT object.
setNumberOfCells(unsigned int value);

testcaseT API

Public Members:

string testNameStr;
string testDescriptionStr;
trafDescTtraf;
benchDescTbench;
rtlLibTrtl;
testMethodsTtestLib;

Public API Functions:

1. Following function is the sole method for generating groups of flows in the stim file. It always returns a reference to a flowDescriptorT object. The return value need not be operated on.   flowDescriptorT objects should never be declared in any other way. If groupSize is not specified, it defaults to 1.
   flowDescriptorT & newFlow(unsigned int groupSize)

2. Following function returns a reference to an object of type negflowDescriptorT. This obejct is of a class derived from flowDescriptorT, so it supports all of flow-DescriptorT's options.
   negflowDescriptorT & newNegFlow(unsigned int groupSize)

3. Following function returns a reference to an rtlFieldsT object with a GLOBAL scope.
   rtlFieldsT & newGlobalRTLConstraint()

4. Following function returns a reference to an rtlFieldsT object with a LOCAL scope. A local scope rtlFieldsT object may be thought of as a flow which has no traffic.
   rtlFieldsT & newLocalRTLConstraint()

5. Following function runs the actual testcaseT object with respect to the device under test.
   Run_Test()

TrafDescT API:

Public API Functions:

1. Following function sets the global default for the number of cells. If this function is not called, the default value is 5 cells generated per flow. Note that the value on a given flow can be overwritten by setting using the flow group's individual setNumberOfCells() function.
   setNumberOfCells(unsigned int value):

2. Following function sets the packet size range for the test. These values default to 8 and 511 bytes respectively if not set. A flow group may override these values by changing its packetSize field.
   setGlobalPacketSize(unsigned int lower, unsigned int upper):

3. These functions set and get the amount of bandwidth available for distribution to the flows programmed into the chip. The input side bandwidth is divided up among all the flows on a per flow basis and is set to the total input capability of the device under test by default. The output side bandwidth is divided among all rateIDs on a per rateID basis. By default this number is set to the maximum available bandwidth usable by the device under test.
   setTotalInputBandwidth(double bitsPerSecond);
   double getTotalInputBandwidth();
   setTotalOutputBandwidth(double bitsPerSecond);
   double getTotalOutputBandwidth();

RtlLibT API:

Public Members:

1. rtlExprLib: This object is a repository for all RTL signal references needed by the stimulus file functions. The API for the rtlExprLibT block is contained within it's <.h> file, but generally, these calls are only needed for complicated benchDescT dependant API calls. In general, most API functions which take these calls also have prettier versions in testMethods which don't operate on HDL variables directly.

2. rtlHardwareDescT: This object holds the actual register model of the device under test. It is public, but contains no actual stuff to twiddle with.

Public API Functions:

1. Following functions get and set the valid range of the FID tags. Since these functions affect the operation and initialization of flowDescriptorT objects in a vary fundamental way, these functions should appear at the very beginning of the user code section of a stimulus file if used at all. Valid values for the maximum and minimum FID are severely constrained by the design of the RTL, these functions do sanity check these inputs.

```
setFIDBase( unsigned int value );
unsigned int getFIDBase();
setFIDMax(unsinged int value );
unsigned int getFIDMax();
```

2. Following functions return the range of valid shaped FIDs in the hardware. There is no set function since these ranges are generally hardwired into the device under test.

```
unsigned int getSMPFIDBase();
unsigned int getSMPFIDMax();
```

3. Following functions set the Scheduler mode for the device under test.

```
setSCHMode( unsigned int mode);
unsigned int getSCHMode();
```

4. Following functions set the Shaper/Meter/Policer mode of the device under test.

```
setSMPMode( unsigned int mode )
unsigned int getSMPMode();
```

5. Following function allocates the amount of bandwidth specified by the value of bitsPerSecond to the specified rateID. Once a rateID is set using this function, it's amount of allocated bandwidth may not be altered by the user.

```
setupRateID(unsigned int rateID, double bitsPerSecond);
```

testMethodsT API:

Public API Functions:

1. Following function creates a local constraint object and calls keepOnly() on it to bind it to the appropriate fields. This function is almost equivalent to declaring a local RTL Constraint and operating on it's respective fields with keepOnly(), but may be faster in some cases. Note that the trafficType field will allow compatible traffic types to appear on that lutPort as per the normal rules.

```
setupLutPort( unsigned int lutPort, unsigned int spiiPort, unsigned int inputInterface, unsigned int trafficType );
```

2. Following function creates a local constraint object and calls keepOnly() on it to bind it to the appropriate fields. This function is almost equivalent to declaring a local RTL Constraint and operating on it's respective fields with keepOnly(), but may be faster in some cases.

```
setupRasPort( unsigned int rasPort, unsigned int spioPort, unsigned int outputInterface, unsigned int trafficType );
```

3. Following function sets up a repeating back Pressure thread task for a given interface and port. For interfaces with no port number, the port number is ignored. percentSlow is a double between 0.0 and 100.0.

setupSlowPortBackPressure(unsigned int inputInterface, unsigned   int   port, double percentSlow);

4. Following function sets up a back Pressure thread task for a given interface and port. The blockage starts at the start of traffic generation and continues for a number of clock cycles specified. The clock signal associated with this function is deduced from the inFace parameter.
   setupBlockageBackPressure(unsigned int inFace, unsigned int port, unsigned int clocks_to_wait);

5. Following function is a more generalized case of the Blockage function, with the start and stop events of the back pressure event controllable by the user.
   setupTriggeredBackPressure(unsigned int _inFace, unsigned int port, tbvEventExprT & start, tbvEventExprT & stop);

6. A blank backPressurePatternT object that may be declared using following function and arbitrary patterns described using the API functions on the raw block.
   backPressurePatternT & getBlankBackPressurePattern();

6.9.1.4 Stimulus file usage

A good stimulus file structure is one that tries to specify the test case requirements in the most direct way possible. Generally, this is also the most efficient way possible. The basic structure of a stim file should be a series of API calls into the Resource Model of the device under test followed by one or more declarations of flowDescriptorT based groups. Here are steps to describe a stimulus for test:

- Step1: Set up the basic structure of the test by declaring a testcaseT object.

- Step 2: Define the Description and testName strings.

- Step 3: Call any testcaseT based API functions which alter the basis settings of the test (particularly functions which affect legal ranges of resources).

- Step 4: Declare any Global RTL Constraints.

- Step 5: Declare a series of Local RTL Constraints

- Step 6: Declare any flowDescriptorT objects or derived class flowDescriptorT objects.

- Step 7: Tell the testcaseT object to Run_Test() on itself.

- Step 8: Declare a tbvExit(); function.

Mostly, steps 3 through 6 can be freely mixed, but it's generally a good practice to separate global scope operations which affect the entire test from local options which affect one or more paths or flow groups.

Using GLOBAL rtlFieldsT objects (newGlobalRTLConstraint()):

In general, most stim files will declare at least one group of flows. Since there are some requirements on all flows or on the device as a whole, these constraints may be efficiently modeled as operations on a rtlFieldsT object with global scope. Since the global scope object of this type are applied to the constraint solver before the local scope objects, these objects can efficiently express constraints that must be true of all flows.

Constraints that are good candidates for a global constraint are those resources which are not exclusive or contended. For example, a global trafficType.keepOut() operation will effectively mask a trafficType from consideration, whereas operating on the trafficType field of every individual flowDescriptorT object will require the randomization engine to solve each flow separately. It is expected that the operations which will be used in the global Constraint space will be: trafficType, inputInterface and outputInterface, since constraining these resources doesn't usually disrupt the other resources from finding valid solutions. Other resource fields such as lutPort, rasPort etc are not so good candidates for a global constraints because they radically alter the available solution space for the device under test.

Declaring Local RTL Constraints vs. Declaring flowDescriptorT's:

In most stim files which involve more than one or two flows, a decision must be made whether to declare some field based resources as flowDescriptorT object or as rtlFieldsT objects. The rule of thumb is to use whichever object can reduce the number of objects created by memory. The initial runtime cost incurred by solving a flowDescriptorT object is the same as that for solving a rtlFieldsT object but the total runtime cost for a flowDescriptorT object is completely dominated by the group size of the flowDescriptorT object.

The flowDescriptorT object will also produce significant runtime cost later when the flows the flowDescriptorT object represents are unpacked. In general, for trivial cases, a flowDescriptorT object will be most efficient since it doesn't require a second object. For tests where the number of objects declared is on the order of 10's of constraint objects, the later runtime saving of not having to program all those flows will become dominant. Since rtlFieldsT object do not unpack into actual routing flows, they are essentially a constant runtime cost item to route.

Also, if you are using localRTL Constraints - the number of overall run length of the test may be easily controlled if the flow group declarations are somewhat separated from the static configuration of the device under test. All field based operations are done using Local RTL Constraints (i.e. RtlFieldT objects with local scope) and the flowDescriptorT objects are only configured to make sure they attach to the right configuration path through the device.

To summarize, the best approach is to use local RTL Constraints for RTL settings, and have the flows grouped into as few, large groups as possible. This improves the efficiency as follows:

- The constraint solver runtime and memory requirements are essentially constant.
- Overall runtime of the test is linearly scalable by controlling the size of the flowDescriptorT group declarations.

6.9.1.5 Summary of Stimulus API

To aid the author of test stimulus, a composite summary of the available API structure should be provided as below:

<u>rtlFieldsT and flowDescriptorT family:</u>
    rtlFieldsT (maybe Global or Local):
//members (support similar tbvSmart API)
      inputInterface
      spiiPort
      lutPort
      rasPort
      spioPort
      outputInterface
      trafficType
      qos
      virtualPort
<u>flowDescriptorT (always local, may represent multiple flows):</u>
//members
      schOn   //deprecated, use set/get functions instead
      shpOn   //deprecated, use set/get functions instead
      rateID
      packetSize
//functions
      setFID(unsigned int fid);
      unsigned int getFID();
      enableScheduler();
      disableScheduler();

```
        setSchedulerEnable( bool enable );
        bool  getSchedulerEnable();
        setSchedulerQOS( unsigned int value );
        unsigned int  getSchedulerQOS();
        enableShaper();
        disableShaper();
        setShaperEnable( bool enable );
        bool  getShaperEnable();
        setRateID( unsigned int _rateID );
        unsigned int  getRateID();
        setNumberOfCells(unsigned int value);
```

testcaseT and members:

//members
```
        testNameStr;
        testDescriptionStr;
        traf;
        bench;
        rtl;
        testLib;
```

//functions
```
        flowDescriptorT & newFlow( unsigned int groupSize );
        negflowDescriptorT & newNegFlow( unsigned int groupSize );
        rtlFieldsT & newGlobalRTLConstraint();
        rtlFieldsT & newLocalRTLConstraint();
        Run_Test();
```

trafDescT (accessed by myTestCaseTObject.traf.memberFunction()) :

//functions
```
        setNumberOfCells( unsigned int value );
        setGlobalPacketSize( unsigned int lower, unsigned int upper);
        setTotalInputBandwidth( double bitsPerSecond );
        double getTotalInputBandwidth();
        setTotalOutputBandwidth( double bitsPerSecond );
        double getTotalOutputBandwidth();
```

rtlLibT (accessed by myTestCaseTObject.rtl.memberFunction()) :

//members
```
        rtlExprLib;
```

//functions
```
        setFIDBase( unsigned int value );
        unsigned int getFIDBase();
        setFIDMax(unsinged int value );
        unsigned int getFIDMax();
```

unsigned int getSMPFIDBase();
unsigned int getSMPFIDMax();
setSCHMode(unsigned int mode);
unsigned int getSCHMode();
setupRateID(unsigned int rateID, double bitsPerSecond);

benchDescT (accessed by myTestCaseTObject.bench.memberFunction()) :

setupLutPort(unsigned int lutPort, unsigned int spiiPort, unsigned int inputInterface, unsigned int trafficType);

setupRasPort(unsigned int rasPort, unsigned int spioPort, unsigned int outputInterface, unsigned int trafficType);

setupSlowPortBackPressure(unsigned int inputInterface, unsigned int port, double percentSlow);

setupBlockageBackPressure(unsigned int inFace, unsigned int port, unsigned int clocks_to_wait);

setupTriggeredBackPressure(unsigned int _inFace, unsigned int port, tbvEventExprT & start, tbvEventExprT & stop);

backPressurePatternT & getBlankBackPressurePattern();

setupInputSpiParityError(unsigned int startTime, unsigned int stopTime);

6.9.1.6 Example Stimulus File

```
include "stimLib.h" //Include library
void tbvMain()
{
     testcaseT sanityTest;
     sanityTest.testNameStr = "sanityMixFlows";
     sanityTest.testDescriptionStr = "sample test that sets up mix of random and directed flows with
specific cases";

     // Define global constraints on all flows to be generated
     rtlFieldsT & sanityGlobal = sanityTest.newGlobalRTLConstraint();
     sanityTest.traf.setNumberOfCells(2000); // Send 200 cells/packets
     sanityTest.traf.setGlobalPacketSize(50,111); // Keep packet size to 50, 111

     // Declare 10 flows in group flowGroup1. Its fields will be randomized
     sanityTest.flowGroup1(10);

     // Declare 11th flow as follows. All the fields are specified
     flowDescriptorT & flowGroup2 = test.newFlow();
     anotherFlow.setFID(99999); // Set flow number
     anotherFlow.setVCI(3);  // Set VCI field
     anotherFlow.inputInterface.keepOnly(constLibT::SPI4);
     anotherFlow.outputInterface.keepOnly(constLibT::SPI4);
     anotherFlow.lutPort.keepOnly(1); // Set port number
     anotherFlow.spiiPort.keepOnly(1);
     anotherFlow.trafficType.keepOnly(9); // Set traffic type to be packet
     // Only turn the scehduler on, no shaping on this flow
     anotherFlow.schOn.keepOnly(0);
     anotherFlow.shpOn.keepOnly(1);
```

```
anotherFlow.packetSize.keepOnly(6);
anotherFlow.qos = 3;


// Declare another group of flow, where need to define only some of the fields.
// Rest are randomized
flowDescriptorT & flowGroup3 = test.newFlow(5);
flowGroup3.inputInterface.keepOnly(constLibT::POS3TX);
flowGroup3.outputInterface.keepOnly(constLibT::POS3RX);
flowGroup3.trafficType.keepOnly(0,6);

// Set backpressure on port 1, SPI4 interface

sanityTest.testLib.setupSlowPortBackPressure(constLibT::SPI4, 1, 40.0);

sanityTest.Run_Test();
tbvOut << "Test " << sanityTest.testNameStr << "complete." << endl;
tbvExit();

}
```

6.10   Monitors

Monitors for each block interface in the datapath and the control path pass on the interface information to checkers for processing. Monitor TVMs pass on the cell received at input and output interfaces to the checker. For the datapath the same class or data structure can be used for all the interfaces, but for the control path, generally every interface between blocks is different. So, each interface needs its own data structure. An example of one such monitor is given in the appendix. Common code between all the monitors can be put in some common functions in the infrastructure instead of duplicating it. Monitor body task should have just the protocol to be monitored.

6.11   Checkers

Checkers collect cells/packets at the input interface generated by traffic generation and transform it to expected output response. This is checked against the actual output response thereby making the verification to be self-checking.

For data path data integrity is verified on all kinds of traffic by expecting a required transformation at the output.

For control path the data coming out can be reordered and this requires the implementation of more sophisticated checkers which predict the right order as well as timings.

It is recommended to cover all possible checkers for the chip by using various concurrency resources supported by testbuilder or systemC or any other HVLs– semaphores, mutexes, barriers, waitChildren(), waitDescendents(), mailBox(), in the environment.

6.11.1 Methodology used in reference design

This section discusses the structure and flow of data path checker for reference design. Also, it highlights the modifications that can be done to enable control path checking for the chip. The checker code is split into common checker code and the block checker code.

tbvChecker(common Checker code) is generic code that has smart queues, ReceiveTask, CheckTask and default process, locate and compare functions. ReceiveTask and Checktask are spawned from the checker that is called from the test. Recievetask tracks the cells coming in the input interface. It puts the cell in the smart queue for cell bursts in tbvChecker.cc. The smart queue selector is in topLevelChecker.cc. Key for top-level checker is "port" on which cell comes. Cell is added in the queue and process function is called from top Level Checker that transforms the cell and push it in expected smart queue. First cell is transferred to intermediate queue based on Flow Id. When a packet is formed, the transformation is done on the packet and the resulting cell is pushed in expected smart queue. Also, checkTask is called when cells start coming out of the output interface. Overall structure of checkTask is same as receive task. This puts actual cell coming out of chip in actual smart Queue for cell bursts. When packet is formed in actual smart Queue, queue selector returns the key for the cell. Locate is called that locates the index of the cell in the expected queue (that the key represents). This index is passed on to compare function. Locate function is for locating the cell to be compared, and compare function does the actual comparison. Locate also compares the actual cell with the expected queue (with same key) – if it's the first cell found the compare function returns the information that the cell is as expected.
If it is nth cell found in queue, compare function reports that cells at index 1 to n-1 are missing cells.
When cell has been compared for expected or missing, they are popped out of the queue. In the end if there are still cells present in the queue – compare returns output queue not empty message for that key.

For any new block checker following is required, these will override the default functions in tbvChecker.cc:

1)   Queue selector
2)   Process function
3)   Compare function
4)   Locate function

Checking explained above is for the top level at input interface of the chip and output interface of the chip and it checks only for data path transformation.
Datapath checking needs to deal with only transformation, but to add control path checking requires knowledge of order and time when the cell is expected to come out., For integrating control path checking in datapath it is essential that datapath transformation is split into two. One for before control path (segmentation) and other after control path (mainly reassembly).
Refer to following figure 6-13, each block to be checked has input and output interface. Checker for the block takes input cell information from input queue, transforms the cell and pushes it in expected queue. It takes output cell information from actual queue and does the comparison. Actual queue is also input for the next interface.

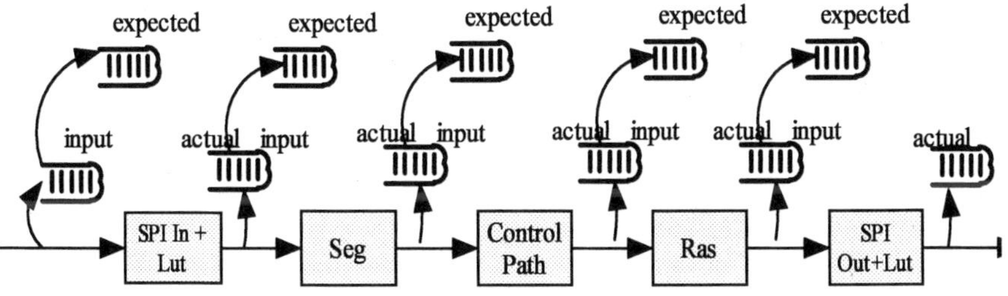

Figure 6-13. Data flow for checker

Here is example structure for how it can be implemented:
cellDispatcherTask: This spawns off monitors at the start of the test. Also, each interface has associated checker and function call (receive Task or check Task) in this code. This way, at the beginning of the test, each monitor has registered checker and rcvTask/chkTask.

CellDispatcherTask.register(blockMonitor – blockChecker and process call) // At the beginning register block monitor with block checker and the function call for process statement.

Now when block monitor is spawned, it knows which checker it is associated with and cell information is copied from cellDispatcherTask to checker input queue. CellDispatcherSpawn task also takes care of deleting this cell when it's passed on to the checker.

Similarly on output side, block output monitor is associated with block checker and chkTask that will call locate and compare functions that transfer the actual queue information to checker actual queue.

A task cellDispatcherSpawnTask in common will be actually spawning all the monitors and associating each of them with checker and function call.

It will pass on the cell information to the checker that is called and then delete it.

```
CellDispatcherTask.register(blockMonitor, blockChecker.receiveTaskCellDispatcherSpawn);
Monitor Spawn(&x)
RegisterTask.blockMon(&x)
Delete (&x);.
```

Flow diagram is in figure 6-14

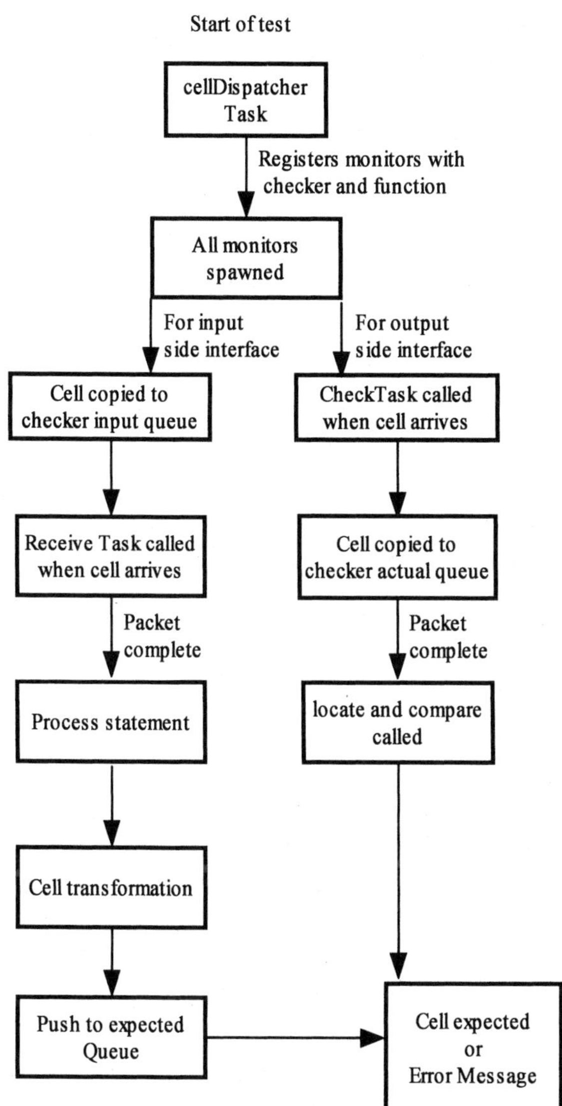

Figure 6-14. Checker Flow

6.11.1.1 Control path checker

Control path checker will either use same approach as above where there are different monitors for all the interfaces inside the control path of the chip. To avoid the monitors at each interface, it can just do checking at control path input and output interfaces.

Above approach for overall checking methodology will need to be altered a little bit to account for control path checking which means for checking any of control blocks functionality like shaping, scheduling or dealing with backpressure in data path. If there is a backpressure, cells keep accumulating in the input queue of the checker. When backpressure is de-asserted this queue has to be flushed immediately. Checker has to know that it has to respond to any event like this immediately to be able to flush the cells when condition is de-asserted. This condition is by default when every cell comes in the interface. To be able to process the cell based on any other event besides incoming cell, checker has to be running independently at the beginning of the test. It will wait for some events based on which it will take an action on the input queue to process the cell. For implementing this, cellDispatcher task will have information for checkers associated with each interface. On seeing a cell, it will push the cell into that checker queue. Now, checker internally will have a function that will keep waiting on some semaphores to check this queue.

Flow: Difference in following flow from data path checker is mainly when some task or function processes the cells at input side interface. Each of the checkers will have its own proccessInput and proccessOutput functions to deal with input and output cells respectively. Celldispatcher task will hook up different monitors with the processInput or output Function of checkers.

Besides processing of cell when a new cell arrives from monitor, checking is also dependent on any other event that demands the cells to be processed before next cell has actually arrived. In this case spawn monitor tasks associated with checkers are simultaneously spawned at the beginning of test with all other monitors. Whenever cell arrives they populate the queue using processInput or processOutput functions of the checker. Also, there can be many other events which require taking some action on the input cell queue. These events will be based on semaphores that get posted when any such cell-processing requirement becomes valid.

This will support following choices in any test:
- Connect single checker for multiple interfaces.

- Multiple checkers for same monitor.
- Use only the monitors, but not checker.
- Start and stop each monitor independently.
- Remove any checker at any time during simulation.

Sample code:

```
spi4MonIn.spi4CtlDatMon.setName("spi4In");
///This adds spiIn monitor, the cells from this monitor is input for topChecker.
cellDispatcher.registerProcessHandleWithMontor(&(spi4MonIn.spi4CtlDatMon), &(tbvCheckerTvmT::pro-
cessInput), topChecker);
spi4MonOut.spi4CtlDatMon.setName("spi4Out");
/// this adds spiOut monitor, the cells from this monitor is for comparison by topChecker.
  cellDispatcher.registerProcessHandleWithMonitor(&(spi4MonOut.spi4CtlDatMon),&(tbvCheckerTvmT::pro-
cessOutput), topChecker);
// lut monitor
  lutSpiiMon.lutSpiiMon.setName("lutMonitor");
/// This just add monitor for spawning.
  cellDispatcher.registerMonitor(&(lutSpiiMon.lutSpiiMon));
```

As shown in flow diagram figure 6-15, in the beginning of a test cellDispatcher task is spawned. This task has all the monitors registered with respective checker functions for input cells or output cells. For example in above sample code, spiIn monitor is attached to process Input of system level checker and spiOut monitor is attached to processOutput function of system level checker. When input monitor sees a cell, it calls processInput function.   Cell is added to input queue. This input queue can be operated on whenever there is arrival of a cell or whenever there is any other events like backpressure or control path signal that needs an action on input cells. The cell is processed based on the action to be taken and pushed to the expected queue. At the same time output monitor is collecting cells in actual queue. It calls locate and compare functions to predict if cell is as expected or not.

CellDispatcher: This task has all the monitors registered to be spawned. Also, it attaches the function pointer to be called when cell arrives. It spawns spawnMonitorTask, which spawns the designated monitor. The main function of this task is create spawnMonitorTaskT object for monitor to be spawned. User can also attach multiple checkers to each monitor. It spawns all the monitors that are attached to it. You can specify monitorTask to be spawned and attach checker function pointer indicating if the cell returned by the monitor is input cell to the checker or output cell to compare.

SpawnMonitor: This task spawns the monitor task and dispatches the cell to checker associated with that monitor

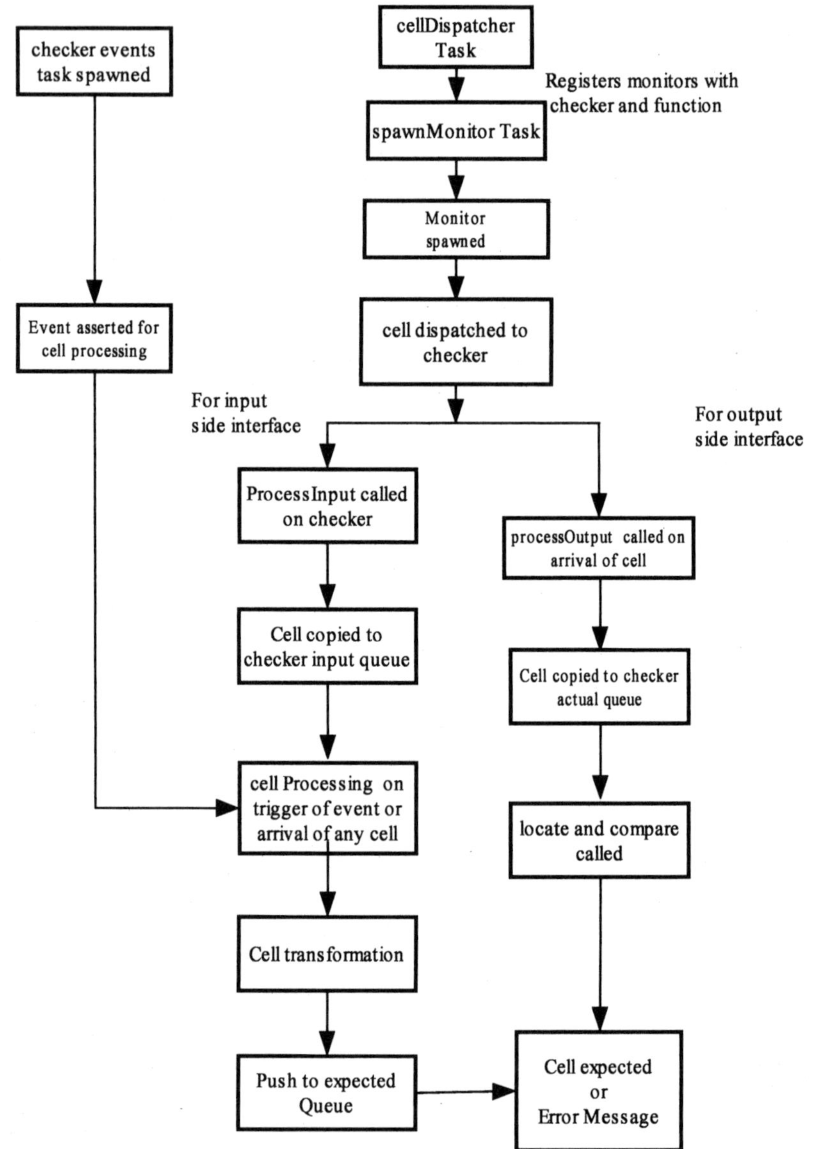

Figure 6-15. Control path checker flow

6.12   Message Responder

During system level simulation, there are multiple threads of generation, monitors and checkers for different interfaces and blocks running. Since there are monitors, checkers, generators, chip configuration setup running at various interfaces simultaneously, each of these need to be able to communicate to central checkpoint where each of these can post the information. For example if monitor at any interface sees unexpected behavior, it can be posted to this central point. This will take an action on these based on the requirement of the system.

This central messaging checkpoint can act on messages from each of the threads at various block interfaces. If monitor returns a CRC error, the information about CRC error has to be communicated to the central messaging checkpoint which at the same time is also receiving any other messages from monitors at other block interfaces in the system. This messaging checkpoint or message handler can decide what action to take if monitor has reported bad CRC error. It might stop the simulation to analyze the problem first or might compare it with the information from other block interface monitors for further processing. It can decide that bad CRC is expected behavior based on the information it has from test stimulus setup. Basically, message handler decides how to process the messages from various monitors, generators or checkers. Here are some typical examples for networking design:

- When chip configuration setup is complete, send an indication to message handler so that it can start the process for verifying that the chip has been setup correctly.

- Monitor can return messages like bad SOP, EOP, CRC error, parity error to message handler.

- When monitor sees an interrupt it sends an info to the message handler to flag a message based on other information it has about rest of the system and also process interrupt registers.

- On receiving backpressure message from one of the monitors, message handler can tell generator to slow down the traffic generation.

- Checkers can send info to message handler on getting some error conditions like if particular type of traffic is getting corrupted, stop generating that traffic type and continue with rest of traffic generation.

6.12.1   Flow

As represented by the flow diagram in figure 6-16, at the beginning of simulation all monitors, checkers, generators and messageResponder are spawned. When a messaging event happens in any of the threads, the information is reported to the message responder. As soon as it is reported, the message is pushed onto the queue. Messageresponder waits on this queue for any message to come and call the corresponding message handler to take appropriate action.

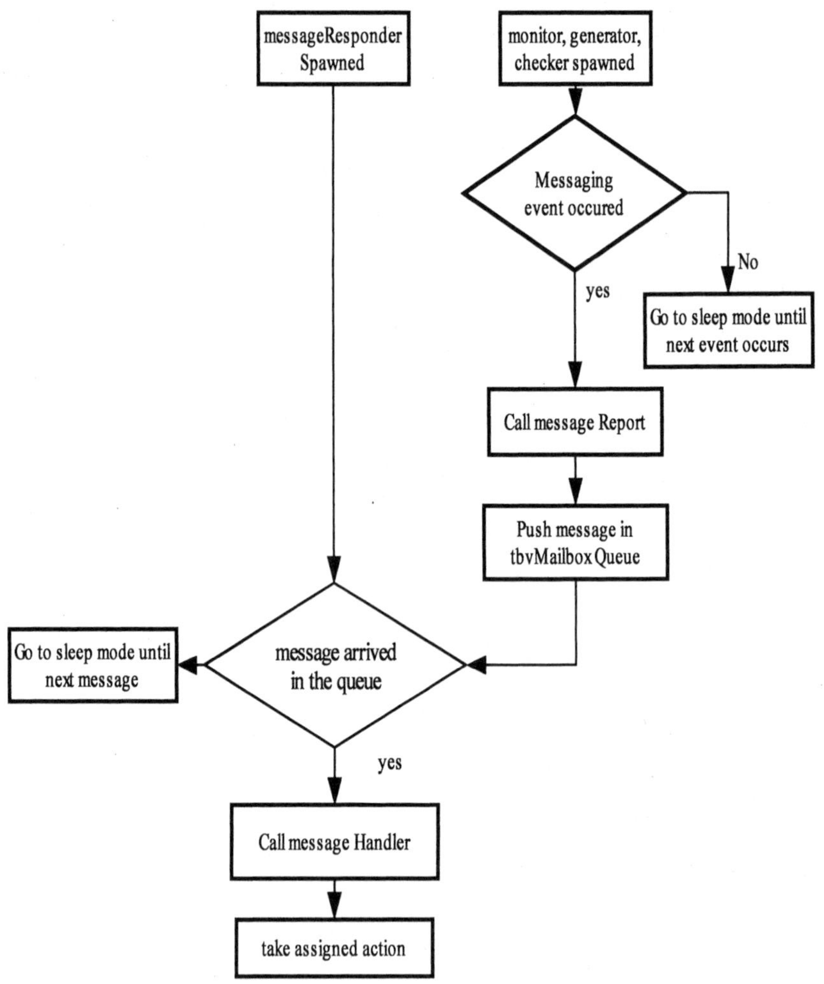

Figure 6-16. Message responder flow

Since various simultaneous threads are trying to talk to report message to the queue, some semaphore mechanism is required where semaphores can be posted for each event. Testbuilder provides a class called tbvMailboxT which can be used as queue of messages. tbvMailboxT class lets you synchronize and pass data between threads in a testbench. It combines the features of queues and semaphores. the tbvMailboxT object starts out empty. You can use tbvMailboxT::push() to add an entry and tbvMailboxT::pop() to remove an entry. The tbvMeailboxT::pop() method suspends if the queue is empty. You can specify maximum size during declaration - if so, tbvMailboxT::push() suspends if the queue is full. Later version of Testbuilder - SCV has sc_fifo and cve_dequeue that can be used.

6.12.2 Basic Structure

Here is the example of basic structure of how message responder can be used for bad CRC example:

```
messageResponderT messageResponder("messageResponder");
///registers message with the message handler
messageResponder.registerMessageWithMessageHandler(messageT::BAD_CRC,badCrcHandler);
messageT & msg = messageResponderT::createMessage();
msg.setMessage(messageT::BAD_CRC);
msg.setUserMessageString("Wrong CRC detected in Ras"); //optional
msg.setData(tbvCell);
messageResponderT::report(msg);
```

Here is the sequence of steps that are executed from above code:

- createMessage() returns reference to messageT.

- messageT has messageType, tbvSmartDataT* and userMessageString.

- setMessage(..) messageType and setData(..) sets tbvSmartDataT *.

- messageResponderT::report(msg) actually reports to messageResponderT that something has happened that needs to be propagated.

- messageResponderT simply waits for any message event to happen and calls spawn on corresponding errorHandler. errorHandler(for example badCrcHandler) is task, which should inherit from messageHandlerT.

- if no error handler is registered it simply logs the message from userMessageString.

- Finally messageT * are deleted once errorHandler is called.

6.12.3 Example code

Following is basic example code. It can be expanded based on the system requirement:

6.12.3.1   messageResponder

messageresponder.h

```
#ifndef MESSAGE_RESPONDER_H
#define MESSAGE_RESPONDER_H
#include "TestBuilder.h"
#include "message.h"
#include "messageHandler.h"
/** \ingroup messagingStructure message responder

This class responds to a message handling request and calls appropriate messageHandler The
message handling requests are posted into a queue, and this task waits on that q, gets the mes-
sage(instance of messageT),looks for appropriate messageHandler(must be sub class of message-
HandlerT) and call run on that message handler.Run must return in 0 time.

*/

class messageResponderT : public tbvTaskT {
public:
 messageResponderT(const char * nameP = NULL);
 ~messageResponderT();
 /// User reports messages using this function
 static void report(tbvSafeHandleT<messageT> msg);
 /// return reference to messageT
 static  tbvSafeHandleT<messageT> createMessage();
 /// register messages with messageHandler instance.
 /// Whenever msg is caught run() will be called on msgHandler

  void  registerMessageWithMessageHandler(messageT::messageType msg , messageHandlerT *
 msgHandler);

 /// spawns this task
 void start();
 /// kills the current thread
 void stop();
protected:
 void body(tbvSmartDataT * dataP = NULL);
private:
 static tbvMailboxT<tbvSafeHandleT<messageT>> * messageQueue;
 map<messageT::messageType,messageHandlerT*> messageTypeToMessageHandler;
  void setup();

};
#endif

/**********************************************************/
```

messageResponder.cc

```
#include "messageResponder.h"
tbvMailboxT<tbvSafeHandleT<messageT>>*messageResponderT::messageQueue = new   tbv-
MailboxT<tbvSafeHandleT<messageT> >("messageQueue");
```

```cpp
messageResponderT::messageResponderT(const char * nameP = NULL){
 if(nameP != NULL)

   setName(nameP);
 setup();
}

messageResponderT::~messageResponderT(){
 delete messageQueue;
}

void messageResponderT::report(tbvSafeHandleT<messageT> msg){
  messageQueue->push(msg);
}

void messageResponderT::body(tbvSmartDataT * dataP = NULL){
  while(1){
    tbvSafeHandleT<messageT> msg;
    messageQueue->pop(msg);
// once msg is obtained get the message handler object for this message and spawn those mes-
sageHandler (e.g. get the object which deals with messageT::LUT_PORT_UNDEF). check whether
a msg Handler is registered with this message else call default message(log message)

    messageHandlerT * msgHandler;
    if(messageTypeToMessageHandler.find(msg->getMessage()) !=
    messageTypeToMessageHandler.end()){
    msgHandler = messageTypeToMessageHandler[msg->getMessage()];
// duplicate pointer to data before calling run(..), since msg
// will be deleted immediately
    msgHandler->spawn(msg);
  }
    else{
//log the message

 tbvOut<<"WARNING"<<msg->getUserMessa eString()<<"Occured"<<endl;

  }
 }
}

void  messageResponderT::registerMessageWithMessageHandler(messageT::messageType  msg,
messageHandlerT * msgHandler){ messageTypeToMessageHandler[msg] = msgHandler;
}


tbvSafeHandleT<messageT> messageResponderT::createMessage(){
  return tbvSafeHandleT<messageT>(new messageT());}

void messageResponderT::start(){
 this->spawn();
}
void messageResponderT::stop(){
 tbvThreadT myThread = tbvThreadT::self();
 myThread.kill();
}
void messageResponderT::setup(){
}
/***********************************************************/
```

6.12.3.2   messageHandler

messageHandler.h

```
#ifndef MESSAGE_HANDLER_H
#define MESSAGE_HANDLER_H

#include "TestBuilder.h"
#include "message.h"

/** defgroup messagingStructure Message Handler

   This is absract base class for all type of message handling
   class. All message handling class must inherit from this
   class making them a spawnable.

*/

class messageHandlerT : public tbvTaskTypeSafeT<messageT> {
public:
 messageHandlerT(const char *nameP = NULL);
 ~messageHandlerT(){}

protected:

 /// The subclass may use this queue.(Most likely for further processing at later time).
 tbvMailboxT<messageT *> msgQueue;

 /// \param msg pointer to messageT.
 void pushMessage(messageT * msg);

 void popMessage(messageT * msg);

 //   virtual void body(messageT * msg){}
};

#endif

/*****************************************************/
```

messageHandler.cc

```
#include "messageHandler.h"

messageHandlerT::messageHandlerT(const char * nameP = NULL){
 if(nameP != NULL)
   setName(nameP);
}


void messageHandlerT::pushMessage(messageT * msg){
 msgQueue.push(msg);
}

void messageHandlerT::popMessage(messageT * msg){
 msgQueue.pop(msg);
}

/*****************************************************/
```

6.12.3.3 badCRCHandler

badCRCHandler.h

```
#ifndef BAD_CRC_HANDLER_H
#define BAD_CRC_HANDLER_H

#include "messageHandler.h"

/** \ingroup messageHandler bad crc handler

    This inherits message handler, and implements handling of bad crc condition in Ras. Depending
on the flows it either logs message or registers a message to exit     the test.

*/


class badCrcHandlerT : public messageHandlerT {
 public:
  badCrcHandlerT(char * nameP = NULL);
  ~badCrcHandlerT(){}

 protected:

  void body(tbvSmartDataT * dataP);
};

#endif
/****************************************************/
```

badCRCHandler.cc

```
#include "badCrcHandler.h"

badCrcHandlerT::badCrcHandlerT(char * nameP = NULL){
 if(nameP != NULL)
   setName(nameP);
}


void badCrcHandlerT::body(tbvSmartDataT * dataP){
 //incomplete implemetation
   tbvOut<<"ERROR: "<<" CRC ERROR"<< endl;
}


/****************************************************/
```

6.12.3.4 message

message.h

```
#ifndef MESSAGE_H
#define MESSAGE_H

#include "TestBuilder.h"
/** \ingroup messagingStructure messages
This class provides place holder for message type and place holder for relevent data to supplied to
message handler.

   Typical usage is:
   messageT & msg = messageResponderT::createMessage();
```

```cpp
    msg.setMessage(messageT::LUT_PORT_UNDEF);
    msg.setUserMessagestring("lut port not initialised");
    msg.setData(tbvCell);
    messageResponderT::report(msg);

*/

class messageT : public tbvSmartRecordT{
 public:
  messageT(const char * nameP = NULL);
  ~messageT();

  /*
  typedef enum{
   PARITY_ERROR,
   BAD_CRC
  } messageType;
  */
  typedef int messageType;
  static const messageType BAD_CRC = 0;

  void setMessage(messageType msg);
  void setData(tbvSmartDataT * data);
  void setUserMessageString(string msg);

  const int getMessage();

  tbvSmartDataT * getData();
  string getUserMessageString();


 protected:

 private:
  messageType message;
  tbvSmartDataT * data;
  string userMessageString;
  void setup();
  void wrapup();
};

#endif

/*****************************************************/
```

message.cc

```cpp
#include "message.h"

messageT::messageT(const char * nameP = NULL):
 tbvSmartRecordT(nameP){
  setup();
}

void messageT::setMessage(messageType msg){
 message = msg;
}

messageT::~messageT(){
 wrapup();
}
void messageT::setData(tbvSmartDataT * data){
```

```
// duplicate data, B'se caller is responsible for deleting
  pointers owned by him.  Basically do not tranfers the ownership.

this->data = data->duplicate();
}

tbvSmartDataT * messageT::getData(){
  return data;
}

const int messageT::getMessage(){
  return message;
}
void messageT::setUserMessageString(string msg){
  userMessageString = msg;
}

string messageT::getUserMessageString(){
  return userMessageString;
}

void messageT::setup(){
}

void messageT::wrapup(){
  delete data;
}
```

6.13   Memory models

Memory models can be either bought from commercial memory modeling companies, memory vendors or built in house. Following details are discussed in memory
modeling chapter 9

- Methodology to wrap up any HDL memory model & convert it to utilize memory dynamically.

- Methodology for converting available memory model to dynamic memory
  model.

- How to create dynamic memory model from Verilog/Vhdl model supplied by
  the memory vendors.

Results from regression using internal models and outside vendor models can be
extremely good in terms of Performance (CPU Time) and memory usage as compared to external available memory models.

6.14   Top Level simulation environment

Top level simulation environment should be set appropriately using the verification infrastructure to be able to run the regression for RTL releases. More details on regression running are discussed in chapter 7

- Regression setup for running regression 24 hours, 7 days a week.
- Script requirements for setting up & running thousands of tests concurrently for months using load sharing software.
- Automatic reporting through email & web.
- Verification metrics for multimillion gate chip in terms of bug rates, exit criteria, version control & coverage.

6.15   Results of using well defined methodology

If well defined methodology is used for verification instead of adhoc quick workarounds for verifying the chip, it pays in long run. You can achieve following goals besides accurately verifying the chip functionality:

- Can achieve successful tapeout with first pass silicon success.
- Can easily find bugs in IPs from vendors that their testbenches could not catch.
- Can have complete verification infrastructure for any networking design using data structure and functionality provided by HVLs.
- Reusable, independent and self checking tests that can drive stimulus independent of the interface.
- TVMs written for block interfaces, traffic generation, block monitors and checkers can be reused with any test environment with different versions of netlist or DUT.
- Robust testbench development that can enable anyone to describe complex random situations in concise format.
- Concurrency features that allow to send in high rate of traffic on various interfaces and perform dynamic result checking and recording.
- Transaction analyses features allow designers to debug much faster.
- Functional coverage for doing post analysis and test bench/regression can help in improvement for next chip.

CHAPTER 7

Regression/Setup and Run

Every time Design increases in size, the number of simulations required to verify the functionality of design increase almost exponentially and so does the challenge of setting up a regression that can lead to preferable exit criteria for signing off on design functionality. Most commonly used metric for design verification is still dependent on number of simulation cycles, since large and complex verification regression can use billions of simulation cycles. So, as design complexity increases we need much faster simulations. Although there has been significant improvement in performance of simulators and workstations, we still need to focus on better regression setup to determine how well verification is progressing and to obtain accurate metrics based on bug rate and test/RTL coverage that can achieve our goal of some pre defined exit criteria.

This chapter focuses on setting up an efficient regression, combining performance of simulators and hardware to verify most complex designs. It illustrates following topics:

- Goals of Regression
- Regression Flow
- Regression Phases
- Regression Components
- Load Sharing

- Features of good Regression
- Common switches for Regression
- Reporting Mechanism
- Exit criteria
- Regression Metrics
- Common Dilemmas in Regression

Most of the above topics are illustrated with an example of running regression on multimillion gate device with first pass silicon success.

7.1   Goals of Regression

Goal of regression is to setup and run the test suite to verify functionality of design and sign off on RTL so that it can be passed on to next process in design phase. Regression has to be run continuously until some pre defined exit criteria for signing off has been met. To achieve this goal regression should be setup so that it can run on the design at any time with minimum effort without any hassles of environment setup and analysis, preferably just by running one command. Whole test suite should be able to run on the design without going through any complex setup process and it should be able to report pass/fail status automatically. With a good regression setup everything can be re-tested whenever need arises and it should be able to run periodically for days or months 24 hours 7 days a week. These are few reasons why regression is needed:

- New version of Design has been released that requires running of same test suite plus added tests for additional features.
- Bug has been fixed in the design that involves major architectural modification.
- Bug fix in the design that appears to be a minor tweak in one block of the design, but has ramifications throughout the other parts of the design when this block interacts with the other blocks.
- Large complex design that has many teams contributing to different modules of the design. One small modification at block level can lead to some impact on the other block a system level that designer is not aware of.
- Some important netlist changes that require same test suite to be run to quickly catch any bug in implementation.
- some new tool installation or modification in verification infrastructure that requires regression to be run to verify that whole verification environment is not effected

7.2   Regression Flow and Phases

Planning for regression starts early in the verification process while verification methodology is being planned that can allow you to run regression on large network of machines simultaneously. Figure 7-1 represents the usual flow for regression:

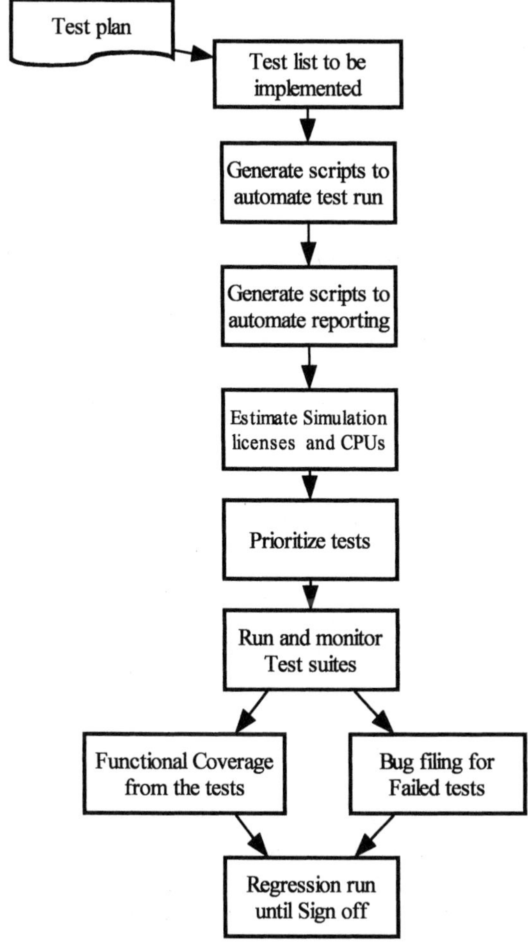

Figure 7-1. Regression flow

Once functional coverage is gathered, the tests need to be re-prioritized based on which ones gives the most coverage and there may be some tests that end up covering nothing new and can be removed from the test suite. Regression runs until all tests all complete and sign off criteria is met.

7.2.1 Regression Phases

Regression running and setup is very time consuming process. Regression suite can grow large enough so that it is not practical to run the whole suite, even on several workstations simultaneously. As modifications are done to the design and regression suite grows, there are more potential problems in the design to be verified.

Figure 7-2 represents the flow for various phases of regression.

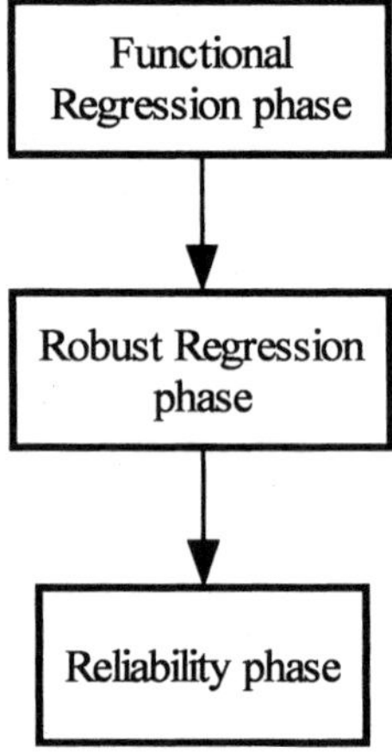

Figure 7-2. Regression phases

A typical design cycle will spend 40% in functional regression phase, 20% in robust phase and 40% in Reliability phase.

Functional Regression Phase:

The goal here to try to cover main functionality of the RTL within a week time that can establish basic functionality of the RTL.

- **Sanity Regression:** The first time RTL is released, a group of tests should be selected and run to check the basic functionality of RTL in two to three hours. Again, regression should be setup in a way that only list file pointing to verilog needs to be modified if at all. Rest of it should be one line command to start the sanity regression. This list of tests can have tests ranging from 1 to 100 based on length of each test or design complexity. These tests are generally repetitive tests run with same seeds to catch any connectivity problems or any other major fallback from previous working functionality because of modifications in RTL.

- **Directed tests:** This Regression suite can be made out of smaller tests. Many Smaller tests can help isolate the problem faster, also it is more stable against simulation crashes because of machine or memory problems associated with larger tests. These are essential specially in finding problems with initial releases of the RTL until it is stabilized to some extent which can be determined by tracking the decrease in bug rate. Several small tests can be set up to run on multiple machines with target execution duration of 72 hours to one week depending on design complexity. Any load sharing tools can be used or different machines can be setup for running different set of tests. These directed tests can try to cover all tests from test plan, randomizing only the specific stimulus in an intelligent manner.

Robust Regression:

After the basic functionality of RTL is proven or validated, next phase is to run more robust tests. These can run from week to a month time-frame and all of the tests are automatic.

- **Random Tests:** Al the stimulus is randomly generated within the constraints of the design. These tests require more simulation cycles, but less effort to produce the tests since they are huge in number. Same test can be run multiple times generating random amount of stimulus to cover all possible ranges of various features provided by the design. In limited time and design process there is no other way to extensively cover in detail all the features with all possible parameters. More reliable random generator tool is required that can allow to generate data randomly and efficiently within the constraints. After running and debugging a preliminary set of manually directed tests, boundary case generation should be utilized to generate the boundary corner case scenarios for one or more random variables. Next, cyclical random generation is utilized to generate the minimum exhaustive set of test parameters. The purpose of using cyclical random test generation is to increase the functional coverage using the smallest amount of simulation. Finally, exhaustive testing is performed utilizing weighted gaussian distribution to generate test scenarios

Reliability phase:

The final phase with running regression is to improve reliability and achieve the goal of exit criteria for RTL sign off. In this phase mainly visibility and monitoring is maximized and directives and switches for all the monitoring is turned on to gather statics for verification metrics that can help management to track the progress. These tests should be able to generate exhaustively all state transitions or stimulus to the design, even those that designer interpreted as not very important things to test. These tests keep running until bug rate has decreased to some particular value or until RTL is signed off and it is time to move on to next process in the design cycle.

7.3   Regression Components

Regression components are as important as any other design components in meeting time-to-market constraints. It is essential to build a sophisticated test suite using pre-existing utilities available. Here is list of some important regression components that help increase productivity of verification engineers that is discussed in following sections.

- bug tracking
- hardware
- host resources
- load sharing software
- regression scripts
- regression test generation
- version control

7.3.1 Bug tracking

There are numerous bug tracking tools available both commercially and as open source. It is a good practice to buy web based bug tracking tool that can be customized. It should be one of the standards in design process to file bugs for any RTL change after it is released. It is one of the best metrics to track verification progress -as bug rate starts decreasing design gets closer to sign off. All failed tests from regression that are result of problem in RTL should be filed as a bug - this can also be automated. Mostly until the regression is stable, it is a good practice to first analyze the error reported to see if it is because of change in regression environment or RTL problem.

7.3.2 Hardware

Simulation runs for large complex designs requires billions of simulation cycles and large amount of hardware resources. While selecting hardware for running robust regressions following points should be considered.

- Should be able to deploy more capacity, supporting simulation runs for months, 7 days a week all 24 hours.
- Should be cost effective and allow maximum utilization of hardware and software resources available.
- Should maximize the utilization of licenses and workstations available for running jobs.
- Should be able to expand the capacity for more simulations if necessary.
- There must be enough RAM for simulation. Disk swapping is known to cause at least 10xdegradation on regression throughput.
- To improve disk access performance, regression should be tried to run locally. If you have too many disk accesses during simulation, it can slow down the performance a lot.
- Several simulation processes can be run on multiprocessor machine. Buying Single multiprocessor machine can be cost effective - it can be used to run various small RTL jobs or one large robust job or gate netlist simulation for utilizing memory space.

7.3.3 Host Resources

Comprehensive resource and load information about all hosts in the network is required for running regression jobs.

Resource information includes the number of processors on each host, total physical memory available to user jobs, the type, model, and relative speed of each host, the special resources available on each host, and the time windows when a host is available for load sharing.

Dynamic load information includes:

- CPU load
- available real memory
- available swap space
- paging activity
- I/O activity
- number of interactive users logged in
- interactive idle time
- space in the /tmp directory

7.3.4 Load Sharing Software

Load sharing software is used to distribute jobs to ensure maximum utilization on available CPUs. Utilities for load sharing can be developed in house or some commercially available utility can be used. These utilities can monitors and control jobs that are submitted through a queue to ensure peak usage of all machines.

Network of heterogeneous computers can be used as single system, regression is no longer limited to resources of one computer from where it is started. Thousands of jobs can be submitted and load sharing software automatically selects hosts in heterogeneous environment based on current load conditions. All jobs are distributed across all nodes with proper priorities, sending results through email, pop-up windows with test pass/fail status, when jobs finish.

There is no need to rewrite or change regression programs, high priority jobs that require fast turn around are assigned to special queues that have access to fastest and least loaded machines, while low priority jobs wait until more machines are available. Batch jobs or remotely run jobs are submitted automatically when required resources become available, or when system is lightly loaded.

There are software available like Grid, LSF etc. for load sharing. Let us take an example of LSF to go into more details of features provided by these tools. They maintains full control over the jobs, including the ability to suspend or resume the jobs based on machine or load conditions. If machine fails than job can be submitted to other machines available in the pool.

With LSF, many engineers can submit different set of tests from regression and the system administrators can easily control access to resources such as:

- who can submit jobs and which hosts they can use
- how many jobs specific users or user groups can run simultaneously
- time windows during which each host can run load shared jobs
- load conditions under which specific hosts can accept jobs or have to suspend some jobs
- resource limits for jobs submitted to specific queues

LSF provides mechanisms for resource and job accounting. This information can help the administrator to find out which applications or users are consuming resources, at what times of the day (or week) the system is busy or idle, and whether certain resources are overloaded and need to be expanded or upgraded.

Multiple queues can be set for running various jobs like system level regression by verification engineers, block level regression by block designers or any other jobs. Each queue has priority based on the policies which decided when to schedule the job.

You only need to learn a few simple commands and the resources of your entire network will be within reach.

7.3.5 Regression Scripts

Regression run scripts are separate from the verification infrastructure of generators, monitors and checkers. These scripts can be developed independent of the verification test suite in any scripting language. Most commonly used language is perl. These scripts target one line command that takes various options for different tools used for running the simulations and parse/analyze log files for reporting. There are options included for enabling/disabling various switches while running regression. For input a list of tests can be taken which are started to run based on machines, licenses and other resources available. Cron jobs are set for automatic running of regressions and for reporting through email or through web based regression status.

7.3.6 Regression Test Generation

Test plan is developed and refined simultaneously with the design phase. All of these different tests make up the test suite. To develop test suite, there is a need to apply stimulus to the design to try to toggle all the signals and also to be able to simulate the response of the system. There is variety of stimulus and responses to simulate like interrupts coming into the system, messages to peripherals, I/O activity and response to the environment. Following is list of various categories of tests generated for regression.

- **Sanity tests:** group of tests to run within couple of hours to pass basic functionality of RTL.

- **Directed tests:** group of tests that run for about of week to cover basic features of RTL to build confidence on main functionality.

- **Robust tests:** group of tests to be run randomly for months on multiple machines to cover most of the feature list. These can be generated automatically using some random generator functions.

- **Reliability tests:** Group of tests to be run to achieve the goal of exit criteria for regression.

7.3.7 Version control

Everything in regression should be under revision control. All designers should follow process of checking in their code whenever it is modified. Once it is ready to be released, it can be tagged and the build should be released to Verification for running regression. On the other hand, verification environment of scripts, models, checkers, monitors, generators etc. should all be using version control. Anyone should be able to check out any past version of RTL and verification to reproduce the problem that has been ever tracked in the bug tracker. The environment has to be tagged properly and current tag for the verification environment that found the bug should be mentioned in the description of the bug.

7.4   Regression features

Following sub sections discuss some important features for most effective and reliable regression and common switches used for running regression.

7.4.1 Important features for regression

- **Separate Verification Team:** Separate verification team should produce system tests because interpretation of spec is done independently. While designer is designing logic to specification, verification engineer is writing tests according to the specification. A consistent interpretation is essential for tests to run on any design.

- **Self Checking:** Tests should be self checking. There can be many techniques used for self checking testbenches based on the availability of tools and infrastructure. Self checking model can do just the consistency check or it can be a complete duplicate model. Generally these models are implemented in higher level verification language. Checker model can be running concurrently and processing data for correctness to flag any error which will stop the test.

- **Easy to use:** Creating new tests should be simple in verification environment. The engineer writing test need not have to understand low level programs, he should have access to high level functions which can be used as test building blocks.

- **Higher level language support:** Verification environment should be able to support high speed behavior models and also switch to more accurate RTL models whenever necessary. Running all the regression in HDL can impact the performance of simulations since most of the systems that are designed these days are large and complex. However, to sign off on RTL, it is required that all regression must be run on full final RTL.

- **Long Lasting Regression** should be able to adapt to any new set of tools, hardware or project. It should not depend on any vendor specific tool and should be able to be expanded when required. It can be very challenging to make regression tool independent because of existence of so many different HVLs and verification tools.

- **Independent of machines:** Regression should be able to run on multiple machines or remotely and should be able to install at any other site.

- **Tool independent:** Methodology followed to build the environment should support usage of different tools or hardware available. Since most large companies have separate design groups having their own preferred list of tools and hardware. Different groups can use different tools/languages but should use the same methodology.

- **Concise format of tests:** Verification infrastructure should be designed such that it should be easy to specify in the test which stimulus data not to randomize and all the rest of essential data should be intelligently randomized. While running the regression if none of the parameters are specified directly, all the data should be able to generate accurately within the constraints. This way every time regression is run, it can generate data whose details are not even specified in the test plan and can expose new problems.

- **Repeatable:** Any random tests or other tests should be repeatable. It is important to specify a seed for the test in order to replicate the error that was found in previous simulation run.

- **Standard:** Regression running should be standard part of design flow and should be supported by the verification environment.

- **Traceability:** Maximum observation and visibility should be provided by the regression. It should be possible to dive in to any interface and extract data from all major interfaces. There should be high traceability into all the models and proper monitoring should happen during simulation run that can save lot of debug time later on if problem is found.

- **Reusable:** All the models and scripts developed during one regression setup for particular project should be reusable in future projects too. The whole environment should be designed to allow as much reusability as it can for various interfaces, stimulus generators, checkers, monitors etc.

- **Well Documented:** To be able to use regression as intellectual property, it is essential that the proper documentation is done and code is structured and well within coding guidelines.

- **Accurate Metrics:** Regression run should be able to generate accurate verification metrics to present clear picture of overall progress and status.

- **Version Control:** Verification infrastructure should be under version control. There should be a process followed to release the right code and to be able to reproduce any problem from previous versions of code.

- **Various Switches:** It should be able to select or deselect directives or switches for saving simulation performance. Mainly switches for debugging, capturing detailed waveforms/database should be avoided during regular regression runs.

- **Controllability:** You should be able to apply stimulus and inject data into any point/interface in the design. If the module is not present it should be bypassed or disabled. Even if some RTL blocks are missing, it should be possible to run simulation at system level either by bypassing these or replacing these with some behavior models.

- **Correct timescale:** Every interface should use the lowest timescale to improve performance. If different interfaces of design have different clock speed - run scripts should be able to alter time scales in the corresponding test cases. For example for SPI4 interface we might need timescale of 1ps but for PL3 interface, 1ns is enough.

  Using this option can increase RTL performance specially when no timing checks are performed. If library elements from vendor contain 'timescale of 1ps/1ps, those library cells cannot be modified, and if simulation is to be run with no timing checks, global timescale of 1ns/1ns can be set -

  - to improve simulation performance

  - to decrease size of waveform database

  and yet retain library cells as they will be used for future gate-level simulations.

Following are the ways to include timescale for all module definitions:

- add `timescale directive to each module of verilog HDL

- include `timescale directive in the first unit that is listed on compile command while other units do not override the timescale.

- use -timescale option in simulation command to give default timescale to modules that do not otherwise have it.

timescales let you use modules that were developed with different time units together in the same simulation. Simulators can simulate a design that contains both a module whose delays are specified in nanoseconds and a module whose delays are specified in picoseconds.

The `timescale compiler directive specifies the unit of measurement for time and delay values, as well as the degree of accuracy, for the delays in all subsequent modules until simulator reads another `timescale compiler directive. This rule applies to modules in subsequent files also. Simulator simulates with one time step for the entire design. The simulation time step is the smallest <time_precision> argument specified in all the `timescale compiler directives in the design. During compilation, simulator converts the delays on accelerated gates to the simulation time unit. This means that the larger the difference between the smallest time precision argument and the typical delay on accelerated gates the more memory your design requires and the slower the simulation speed. Therefore, you should make your <time_precision> arguments no smaller than is necessary.

'timescale <time_unit> / <time_precision>

The <time_unit> argument specifies the unit of measurement for times and delays. The <time_precision> argument specifies the degree of precision to which simulator rounds delay values before using them in simulation. The values simulator uses will be accurate to within the unit of time that you specify. The smallest <time_precision> argument of all the `timescale compiler directives in the design determines the time step for the simulation.

The <time_precision> argument must be at least as precise as the <time_unit> argument; it cannot specify a longer unit of time than <time_unit>.

So, there must be a `timescale directive in front of every module that has #delays inside the module. Because, A `timescale directive is a compiler directive that is active for every module compiled after the directive until another `timescale directive is read. Adding a `timescale to every module that has delays insures that the correctly scaled delays are used during simulation. If there are delays in a module and if the wrong `timescale is active, the delays will be scaled wrong by an order of magnitude or more.

For interpreted simulators such as Verilog-XL, using timescale of 1ns/1ps requires about 56% more mem and about 99% more time to simulate than the same models using timescale of 1ns/1ns

- **Option for generating tool independent test data:** Everyone's verification environment is dependent on tools and languages they selected for the verification infrastructure implementation. There should be a way to convert the current environment to be used by any customer with any set of tools. One possibility is to dump the vectors during regression run and have ability to play them back.

7.4.2 Common switches for Regression

Every time regression is run, there might not be everything that needs to be enabled for example, detailed monitoring or full hierarchical recording might not be required during initial runs. This can save a lot in terms of performance. Only when HDL or verification has to be debugged for detailed analysis some of these switches can be used. This section discusses switches mostly for regression:

- What should be limits on database recording. The waveform size can go on increasing if you run test for longer time and have enabled recording of all the signals.Waveform size and what signals to record is one of the common switches used. When this waveform size is reached, it is cutoff from the zero time and records only latest results.

- Whether to always record waveform or only on test failure. At the stage when RTL is quite stable, the tests can be run with logging out lot of relevant information and rerun to generate waveform only in case of failure and further debugging. Else, from thousands of tests if all waveforms are saved even if there is nothing to analyze, it can consume lot of disk space.

- How many checkers to include. It might hit the performance if all the checkers for all blocks and system level are enabled because there will be lot of simultaneous threads running during simulation.

- What interfaces to monitor for analysis. Monitors for rest of the interfaces should be switched off. It will save both disk space and performance.

- Specify which blocks are accurate RTL blocks and which ones are bypass models for continuing simulation at system level.

- What are the versions of RTL release, various tools, memory models or any other s/w or verification infrastructure being used.

- If simulation is being run on gate level netlist, RTL or high level models of the blocks. Initially for developing verification infrastructure only bypass models can be used if block RTL is not ready.

- List of tests to run.

- Whether to do compilation, elaboration or simulation. Initially when RTL is released it has to be compiled and elaborated. Next, when regression is run it only has to be simulated unless RTL is modified or needs to be recompiled for debugging and/or coverage.

- Whether to switch on the coverage or not.

- If any debug options for RTL or verification infrastructure has to be enabled. Debugging can slow down simulations a lot.

- What all reports to be generated from regression and who all should they be sent to automatically.

- At what interfaces to generate stimulus or inject data.

- Whether to turn on performance monitoring for regression. This is to track how fast jobs progress and how much is the memory usage.

- To how much detail progress in simulation is to be traced.

- When to stop the test. Test can be stopped if specified number of errors have been seen already or if specified amount of data has been processed.

7.5   Reporting Mechanism

When an anomaly or error is detected in the test there should be enough of additional information to allow you to zero-in on the problem. Results can be gathered at central location for reviewing and archiving. Several reports are required to be generated from different phases of regression for different audience. The results of regression can be quite voluminous. These are required to be saved for several months. So, analyzing these results can be very time consuming. Figure 7-3 represents generally followed flow for reporting and analyzing results:

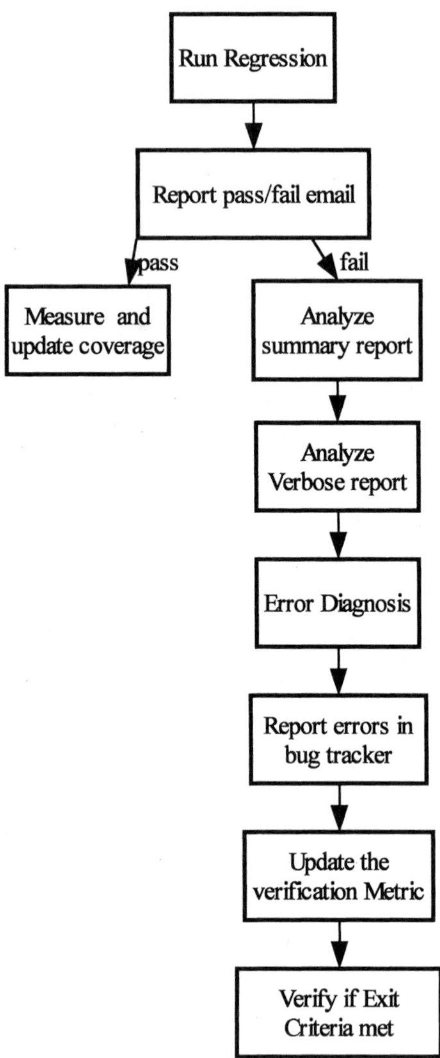

Figure 7-3. Regression Reporting

7.5.1 Pass/fail count Report

Regression is generally run 24 hours, 7 days a week. It is useful to send pass/fail count report with list of tests that failed as email to the verification engineers. It can help to find out quickly if last modification broke something.

7.5.2 Summary for each test case

Summary for each test case should be stored which can help identify if test ever run. What all random parameters changed from last run and how much data was processed. Did it fail during compilation or simulation. Was failure because of license expiration, machine issue, tool, verification environment issue or actual RTL problem. Regression can also fail because of un predicted tool, license expiration or machine issue too and it is a good idea to have one line summary in the test case report that specify these failures. It can also be included in overall summary the first error that caused the test failure.

7.5.3 Verbose Report for Debugging if error occurs

A verbose report is essential for debugging. There should be enough information printed from the interface monitors that can help resolve the errors faster. A fair amount of intermediate data is put out in these reports, and this output from test is used to determine what went wrong. It can either have high level functionality failure listed from checkers or the signal or register values that mismatched. At least, all of these have exact simulation time information narrowed down to the module where the failure occurred.

7.5.4 Error diagnosis

Once test reports an error, it has to be diagnosed and bug has to be reported. If above report files do not give enough of information to debug the problem the test might have to be run to generate detailed database by turning the switches on for displaying and monitoring various nets and registers. In general there are thousands of nets and registers in the design and it is not possible to save all nets and registers transition in log files or waveform database for huge regression every time it is run. Error diagnosis can be complex process. Usually, every morning verification engineers get a report for failures from regression and they spend day in figuring out the details of error while regression is running 24 hours.

7.5.5 Exit criteria

It is challenging to determine that the design is logically correct and it has been simulated enough. Usually regression should be stopped based on some exit criteria and not because there is no more time left. Verification metrics are very helpful in defining a suitable exit criteria for the management.

7.5.6 Verification Metrics

* **Coverage:** During several regression runs, design coverage is monitored for all states, signal transitions to find out if all possible options, fields, rates, ranges and cases have been tested. Branch coverage, expression coverage and state coverage is mostly used to identify blocks that are not adequately verified. Besides doing HDL coverage, it is essential to do functional coverage that determines how much of system functionality has been tested based on the specs. There are several tools available for functional coverage on regression.

* **Monitoring:** Regression should be able to monitor all different cases that have been so far covered and parameters that have been set randomly. This can also help you generate most effective and efficient regression test suite. Monitors at various interfaces ensure maximum observability for debugging or tracking the progress.

* **Bug rate:** Most bug tracking tools are developed to operate on web browser. They give good visual displays in terms or charts, graphs, histograms for current or past project statistics or trends. If bug rate does not decreases to a certain number in period of time, it is decided to move on to next process in the design flow and sign off on verification of RTL. Bug rate gives a clear picture of priority of the bugs found on various features of the chip that helps management to decide to move on to meet the schedule.

7.6 Common Dilemmas in Regression

Here are most common dilemmas associated with the regression

* Running regression is tedious process and to be able to setup and track all the test results and metrics and deal with all hardware/tool issues is complex.

* How often regression should be run. If it is run quite often, it can be utilizing too many resources. If it is not to be run often enough, it might not be achieving enough.

* It is difficult to cut the line between just isolating and tracking the defects versus spending time on debugging them.

* What is the best methodology for getting the test coverage and minimizing the number of tests and get maximum coverage with limited simulation time.

- To keep up with the pressure to sign off on the chip and pressure to keep testing and finding the bugs.

- To be able to make progress in the testing as a whole versus improving quality of current test suite in terms of nailing down the cause of defect and being able to reproduce the bug.

- To reproduce the bug found from long complex test and compose it into smaller test for debugging.

- To draw a line between block level and system level testing.

- Last but not the least - when to stop the regression?

7.7   Example of Regression run

As an example of typical regression run for multimillion gate chip, here are different steps:

- During first release of RTL sanity regression is run which generally is few hours.

- Based on passed status from sanity regression, next test categories are run.

- If all the tests have been run once and this is incremental RTL releases than first category of tests to be run are the ones from the feature that has been added or modified.

- After the added or modified feature is tested rest of the tests can be submitted.

7.7.1   Run options.

Run scripts can be written in any scripting language mainly in Perl and Python. It is also a good idea to have Makefiles for regression run. As more and more features are added to regression scripts, if perl scripts are not written properly it becomes difficult to maintain and add more features to it. Another alternative is to use Makefiles for regression run too. It becomes simpler and more organized to have all the run options and regression run commands in form of Makefiles. With Makefiles, you can give various options and switches for the command line arguments for the simulator and other tools. Small scripts can be written for choosing run options and reporting. What these scripts and Makefiles essentially do is:

- Selection of test or list of tests to be run. There should be an option to run them all together on different CPUs or sequentially one after another

- To pass on the right arguments to the compilers, simulators and other verification tools to be used

- To have an option to compile just the testbench, just the RTL, both testbench and RTL together, and option to only simulate the current test using precompiled library for both RTL and testbench

- Good reporting mechanism - generally there should be reporting scripts running in the background using cron jobs that check for the regression jobs in LSF queues and able to send the current summary status on the jobs that are running or the jobs that completed.

- Reporting scripts should be able to send email to the list of people for the current status on regression run

- The test plan document should be automatically updated from these scripts for the pass, fail status

- Besides reporting using cron jobs, there should be an option to extract the current status anytime and generate all the reports.

Here is one of the examples of the daily summary report generated for the tests that are running in regression.

Priority	Pass	Fail	Not yet run	Not yet implemented	Total	%age run
0	566	0	4	1	571	99%
1	189	4	0	0	193	100%
2	347	0	15	0	362	96%
3	195	2	0	0	197	100%
4	37	0	0	3	40	93%
5	79	4	4	0	87	95%
Total	1413	10	23	4	1450	98%

Besides the summary report, the detailed test document should also be updated by regression reporting scripts.

Besides summary report, a bug chart as shown in figure 7-4 is also needed to be updated after the failed tests are analyses. There are automated tools to do this or scripts can be written to parse log files, summary, previous status and generate status as below which is bug trend for the Design under verification.

Rev	3.1	3.2	3.5	3.8	4	4.1	4.2
Date	15-Dec	25-Dec	2-Jan	9-Jan	16-Jan	23-Jan	30-Jan
Bugs	31	28	43	30	16	16	19

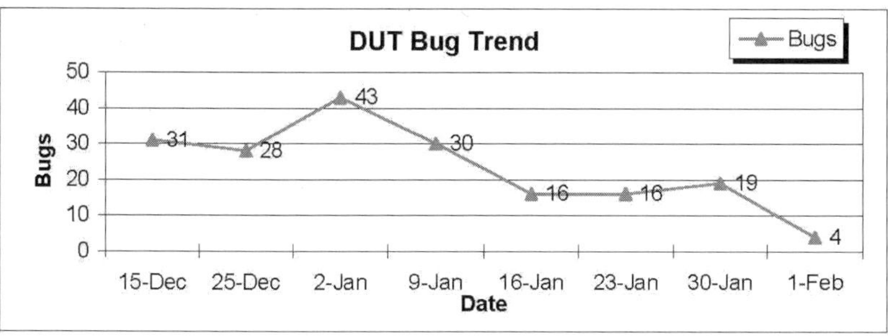

Figure 7-4. Bug chart

7.8   Summary

This chapter highlights all the essential features of automated regression setup and focuses on various trends for running regression, how to setup regression and maintain regression based on your design requirements. Regression running should be an automated process where you start regression from start button and after that daily emails and report is generated until all regression is complete. Usually, this is not the case because there are many tools, jobs, memory space, machine issues which come in the way. So, it is required to do an estimation first on your resources for licenses and CPU slots and setup the regression based on design and functional coverage needs.

CHAPTER 8

Functional Coverage

Functional coverage is a relatively new term that is appearing in the Electronics design industry. What is functional coverage and why is it becoming so important? Traditionally we would all instinctively build functional coverage into our verification environments without thinking about it. Verification plans would include a set of tests that exercises all of the functionality defined in the specification that we would receive. If the tests passed then all the functionality is covered. Our functional coverage metric would be 100%, even though we did not know to call it a functional coverage metric. What has changed is that we can no longer write all the possible tests necessary to prove that 100% of the functionality is exercised. This would require too many tests to write and too much simulation time to execute. Therefore we rely more and more on constrained random testing which allows tests to be directed at exercising functionality while randomizing some of the data in the tests. Since data is randomized there it is no longer a guarantee that a test will exercise a specific function or functions. This creates the need to monitor activity during simulation and determine if the function was exercise.

This chapter addresses the following topics:

- Functional Coverage definition.
- Coverage tools that are used today and how they are different from functional coverage.
- Functional coverage implementation.

- Functional Coverage examples.
- Limitations of Functional Coverage.

8.1   Use of Functional Coverage.

The use of functional coverage is discussed in many different companies and industry forums. The definition of functional coverage used in those discussions, though, is not always clear. This can lead to misunderstanding of what needs to be done as well as what is already being done in this area of verification. This section will give a definition of functional coverage and attempt to answer the question of why should it be used.

8.1.1   Basic Definition Functional Coverage.

Functional Coverage is the determination of how much functionality of a design has been exercised by a verification environment. It requires the development of a list of functionality to be checked, the collection of data that shows the functionality of concern being exercised, and the analysis of the collected data. Functional coverage does not prove that the design properly executed the function, just that it was exercised. It is the job of the testbench to prove proper execution.

Functional coverage is indirectly employed today. For example, if your management told you to prove that the set of regression tests created for a particular design was exercising all the functionality defined in the specification, how would you go about proving it? First of all you would need to show the list of tests in the regression suite and the correlation of those tests to the functionality defined in the specification. Next you would need to prove that the tests actually executed the functionality it is supposed to check. Finally you would create a list showing each function and check off those that were exercised. From this list you would extract a metric showing the number of functions exercised divided by the total number of functions to be checked. This is probably what you would present to management. This is functional coverage. The difficulty is that it is too much of a manual process. In addition, the methods for creating the list of functionality, collecting that data and correlating it to the functionality, is done in an ad-hoc fashion. Today's designs require a more structured and automated approach.

8.1.2   Why Use a Functional Coverage?

There are two magical questions every design team must ask and answer: Is my chip functioning properly? Am I done verifying the chip? The first question relates to the checking of the simulation results. The second question relates to a metric that is used to determine when the verification effort is complete. Functional coverage metrics are a comprehensive way to determine that all the functionality of design has been verified.

There are different ways that projects implement functional coverage from the execution of the test suite:

- Simply verify that each test executed properly.
- Apply a functional coverage tool to create a metric.

Proper execution of each test in a test suite is a natural measure of functional coverage. Each test is created to exercise a particular functionality item in a specification. Therefore it is natural to assume that if it were proven that each test completed properly then the entire set of functionality is exercised. This assumes that each test has been verified to exercise the functionality for which it was created. In many cases this verification was performed manually. The waveforms from a simulation run are viewed by an engineer and the conditions for which the test was written to verify proved to exist. This type of manual checking is both time consuming and error prone. It is usually only performed once or twice, typically when the test is first written. After that time it is assumed that the tests will always exercise the same functionality. Since designs change during a design cycle this assumption may not always be valid. Functionality may be modified during the cycle due to shift in the market the device is being created for. Another possible change relates to protocols that may require an update during the cycle if they are still being evolved, such as PCI-Express, or a new variant starts becoming dominant, such as the 802.a ethernet protocol quickly went to 802.b.

Directed random testing, which is becoming a necessity in today's complex chips, invalidates the use of manual checking. In directed random testing, each test is targeted at a range of functionality while letting the data be randomly selected. For example if the multiply instruction of a processor were to be tested, the opcode for the instruction must be a specific value (i.e. directed). The values for the operands can be any integer. Therefore they can be randomly selected. The value of randomizing this data is to prove that an error does not occur across ranges of values. So while a test can be classified as exercising the multiply instruction, how do you know you have exercised a sufficient number of unique data values to make sure they all work? This is where functional coverage can help.

8.1.2.1 Functional Coverage Flow

The flow of functional coverage should be as follows:

- Determine what functionality you need to prove that it has been exercised.
- Create a Coverage Model for each functionality you have. The Coverage Model includes:
  - Defining coverage items.
  - Creating coverage monitors
  - Creating coverage goals.
- Instrument your environment to capture information that proves execution of functionality.
- Execute your verification suite of tests.
- Collect and analyze the information collected during execution.
- Create a metric from the analysis of the data.

8.2   Using Functional Coverage in Verification environment.

Some form of checking for functionality is done in every verification environment today. Some of the coverage is inherent in the tests themselves. Some teams use code coverage to enhance the checking. Others get even more sophisticated. There are many ways in which verification engineers capture functional information today. Most prevalent is writing information to a log file.

8.2.1   Difference between Functional Coverage and Code Coverage.

There appears to be a confusion in the industry on what constitutes functional coverage. Commercial tools known as code coverage tools have been around for a number of years and have been used on many projects. A natural assumption is that code coverage and functional coverage are one in the same. While they have a similar purpose, which is to determine what part of the design has been exercised, their approach is much different. Code coverage in its simplest form determines which lines of code in the design have been exercised. Tools typically generate a lot of output. While this output has value it is not "actionable" by itself. For example if 10% of the code in the interrupt section has not been exercised, what "action" should be taken? Write more tests that exercise the interrupt logic? What functionality was not exercised? It is difficult to correlate identified holes in code coverage to un-exercised functionality. This shortcoming is efficiently addressed by functional coverage by explicitly measuring the function that has been exercised as opposed to measuring lines of code exercised.

An example can be used to discuss the difference between functional coverage and code coverage relates to word processors. Lets say you need to write a progress report for your management. Once the initial report is written you would normally run a spell check program to make sure there are no spelling mistakes. The spell check program equivalent to basic code coverage. It is easy to implement because all it needs to do is check each word against a dictionary or check that each line is executed. It is necessary because you do not want to send a report to your management that has mis-spelled words, just like you would not want to ship a design in which the interrupt code was never executed. But is that sufficient? Would you ever think about sending out a report without re-reading what you have written to make sure it is coherent or make sure that it properly conveys your ideas? Probably not. Spell checking is not sufficient because it does not address the reason why you wrote the report, only the mechanics or structure, which is the correct spelling of the words. Most modern word processors have grammar check capabilities. A grammar check is a step closer to a functional check. It looks at a series of words to determine if they make grammatical sense. This is analogous to more advanced code coverage techniques such as branch coverage, path coverage, etc. They look at a series of lines of code and try to determine if appropriate behavior is exercised. While grammatical checks help clarify sections of a line, it still does not determine that your ideas are properly articulated in the words. The only way to determine this today is to re-read your document many times or have colleagues read the document to see if the ideas are clear. The same need is there in the design world when trying to determine that all the defined functionality of the design has been exercised. It requires more knowledge of the functionality than counting lines of code executed or states in a state machine executed.

Therefore functional coverage looks at how the functionality of the design is exercised whereby code coverage looks at how low level items of a design are exercised, such as lines of code, state machines, etc. While it is good to use code coverage, it is not sufficient, in the same way that it is not sufficient to only use a spell checker to determine if a document is well written.

8.3   Implementation and Examples of Functional Coverage.

Functional coverage is divided up into three areas: creating the coverage model, capturing functional information, analyzing functional information. But first, let us look at some basic definitions:

- Coverage Model: "a family of coverage events that share common properties are grouped together to form a coverage model".

- Coverage Tasks: Members of the Coverage Model are called coverage tasks and are considered part of the test-plan. This is also commonly referred to as coverage -items.
- Cross Product coverage models: These are defined by a basic event and a set of parameters or attributes, where the list of coverage tasks comprises all possible combinations of values for the attributes.
- Holes: Larger uncovered spaces (coverage = 0)

8.3.1 Coverage Model Design.

Functional coverage models are typically composed of four components. They are:

- Coverage groups: a group of coverage items.
- Coverage Items: data items needed to determine completeness of execution.
- Sampling: determining when to capture information for coverage.
- Coverage Goals: Expectations to measure execution completeness against.

The determination of what functionality to create a coverage model for, involves examining the functional spec and create a list of important functionality that needs to be verified. This list contains attributes of the design that need to be verified and are important. These attributes can be structural, or functional in nature, or can be important from an implementation point of view.

Let us now take a closer look at the individual components of a functional coverage model.

8.3.1.1 Cover Groups

Coverage models are typically created by first isolating attributes of the design that are of interest and then collecting them in groups with some common property or feature that bind them together. This forms the basis of cover groups. A coverage group includes a sampling event and a set of coverage expressions.

8.3.1.2 Coverage Items

Each individual attribute of the design is captured using a coverage expression. These coverage expressions which specify the values of interests of the sampled variables, illegal values, and ignored value, typically represent individual coverage items belonging in the cover group.

Illegal value specifications are used to directly identify bugs more so than provide coverage feedback. Illegal values should never occur and are logged as verification errors. Ignored values define states that are not collected and are ignored while determining the coverage amount.

The individual coverage items determine the "what" information should be captured.

8.3.1.3 Sampling Event.

The next part of the puzzle is the "when", which is typically called the sampling event. This can be as simple as the rising edge of some signal in the HDL, or could be as complex as a complex temporal expression enumerated in both logic and time.

A sampling event is typically an HDL signal in the DUV or variable change in the testbench. The coverage database is updated when a relevant sampling event is triggered.

8.3.1.4 Coverage Goals.

Once you have defined the metrics for coverage collection, then you need to figure out how to quantify and measure the coverage values. For coverage items that has a small range of values, each value needs to be used for the coverage to be considered complete. For items that span across a large range, say a memory, or range of addresses, it is prudent to specify certain sub-ranges. This is a concept called binning. It involves the creation of individual bins over sub-ranges of a large range of values. In this case each bin must have been used for coverage to be considered complete.

Let us look at an example, say a bus with a range of addresses. Let us say DMA operation occurs within a sub-range 0x1F to 0x5F, then we could create a DMA-bin covering the above address range. Alternatively, one could create N number of bins between the above address range, say if for example DMA to DRAM occurs between 0x1F and 0x2F, and DMA to EMIF occurs between 0x3F and 0x4F, and so on.

Coverage Goals are quantitative and numerical targets set as part of the test plan. The goals are usually used as an exit criteria for testing. They are pre-set targets that measure the quality of verification and need to be carefully chosen. Once these targets are reached during simulation, there is high confidence that the corresponding functionality has ben adequately tested. However, depending on achieved code-coverage and assertion-coverage, these numbers are often revised and fine tuned during the course of a project to give a better measure of progress.

8.3.1.5 Coverage Monitor.

During the process of simulation, achieved coverage is constantly monitored to keep track of the progress of simulation. Each individual coverage item is evaluated at the specified sampling event and the coverage count incremented as specified. While in some commercial tools coverage results can be accessed at run-time, in others it can be accessed only at end of simulation.

8.3.1.6 Types of coverage Models.

Coverage models are typically classified into three different categories:

- **Black-Box** - in which case the overage monitors are placed at the interface.
- **White-Box** - in which case the Coverage monitors probe and monitor internal nodes.
- **Grey-Box** - in which case monitors probe both internal nodes and interfaces.

Further classification is based on Temporal properties:

- **Snapshot** - a single point in time. For example, if two routines are called sequentially, a model containing the values of variables from both routines is considered a snapshot model even though the variables are not active at the same time.
- **Temporal** Coverage Models - coverage is monitored on sequence of events, time between events, or on the interactions between events.
- **Stress** Coverage Model - checks how many times properties are evaluated in a short interval of time, or concurrently. The main purpose is to ensure that HDL resources have been fully utilized and tested.
- **Statistical** Coverage Model - The main motivation is collection of statistics - like how many times some events happen.
- Another major kind of coverage model computes the cross product - In a **cross-product** functional coverage model, the list of coverage tasks comprises all possible combinations of values for a given set of attributes.

8.3.1.7 Interpretation of Coverage results.

Once you have quantized the available data, the next step is to interpret the collected data. To start with, if a coverage item has not been exercised at all, which means it has a count of zero, then it is considered a coverage hole. This points to missing stimulus, or the fact that the applied stimulus failed to excite the corresponding circuitry and failed to produce the desired behavior. If the stimulus was generated using constrained-randomization, then those constraints have failed to produce the desired effect. This is a key discovery in the verification process. The verification person should now attempt to either modify the constraints on the generation process or modify the random-number seed for the generation of the corresponding stimulus. The verification of the corresponding section of the DUV is considered incomplete until the coverage goal is met.

For a very detailed, and mathematical analysis of holes for functional coverage, the reader is referred to [Ref: Hole Analysis for Functional Coverage data - Oded Lachish, Eitan Marcus, Shmuel Ur, Avi Ziv of IBM Research lab, Haifa - http://www.research.ibm.com/pics/verification/ps/holes_dac02.pdf]

Another useful concept used in interpretation of results is cross-coverage. This is generated by computing the cross product between two or more coverage items. More on this later.

8.3.2 Transaction-based Functional Coverage techniques.

Traditional coverage techniques are not up to the task of determining the functionality exercised by a constrained, random test environment. Traditional techniques have relied on directed tests. If the tests passed, the functionality was exercised. With constrained random tests, data is randomly generated making it difficult to know specifically what will be tested beforehand. New techniques have been created.

Transaction-based functional coverage is becoming a reality with advent of the transaction recording and call-back schemes inherent in SCV.

The flow of using the recording capabilities of SCV is relatively straightforward. Statements are simply added around current code. Three types of statements can be added: database creation, stream creation, and transaction creation statements.

Database creation statements are self-explanatory. These statements open a database that transaction results are written to. The SCV standard defines a text file for this database. Commercial tools such as Cadence verification tools create a database in their SimVision environment. An example of this code is:

```
scv_tr_db ("database");
```

Stream statements define an object on which to attach transactions. These streams are needed because a transaction is an abstraction that has no hardware equivalent and therefore does not have a physical place in the design that it attaches to. Contrast that to a signal, a more familiar item to design and verification engineers, which is a real piece of hardware. It has a hierarchical path in the design and can be accessed or displayed very easily. Streams were created to give a user the flexibility to define how to present this abstract information. Take a memory controller, for example. Lets say the controller has the following signals: select, read, write, reset, address, data. It is very natural for an engineer to display this information in a waveform tool. What about the following transactions: read, write, reset. These transactions indicate operations that are derived from the signal activity. How do you display this information in a waveform tool? Should all the reads be displayed on a single line like the a signal, and the writes on a different line, and the resets on yet another line? Should all three types of transactions be displayed on the same line? The user decides this by creating a stream and determining which transaction types will go onto that stream.

The format for creating a stream is: scv_tr_stream name ("name");

Transaction creation statements define the transactions that will be created and attached to a particular stream. The format for these statements is: scv_tr_gen <> read ("read");

Adding transactions to existing code is straightforward. Applying these concepts to the transactor code is defined here:

```
SC_MODULE (top) {
  void test () {
    scv_tr_text_init();
    scv_tr_db mydb ("mydb");
    scv_tr_stream bus_stream ("bus_stream");
    scv_tr_gen<> write_gen("write");
    scv_tr_gen<> read_gen("read");

    ...

    For (int I = 0; I < 10; I++) {
    Cout << "data: " << atm << endl;
    arg.next();
    if (arg.r_wn == 1) {
       h1 = read_gen.begin_transaction();
       read (arg);
       read_gen.end_transactoin(h1);
    } else {
       h2 = write_gen.begin_transaction();
```

```
   read (arg);
     write_gen.end_transaction(h2);
  }
   }
 }
```

This code writes a transaction record to the database, in this case a text file, for every write and every read that occurs during simulation. Notice that the SCV tools automatically add the fields of the args data structure into the transaction. This information now can be used to measure functional coverage.

Transactions created can be used on-the-fly for functional coverage analysis as well as post- processing functional coverage analysis.

8.3.3 Functional Coverage Examples

8.3.3.1 Video systems

The example circuit is a wireless digital picture frame. The device takes an encrypted jpeg video image via one of its ethernet ports, decrypts the image, then decompresses the image, and finally sends the result out its VGA display port. The block diagram of the circuit is shown below in figure 8-1. There are four major blocks, Ethernet, DES, JPEG, VGA, plus a controller.

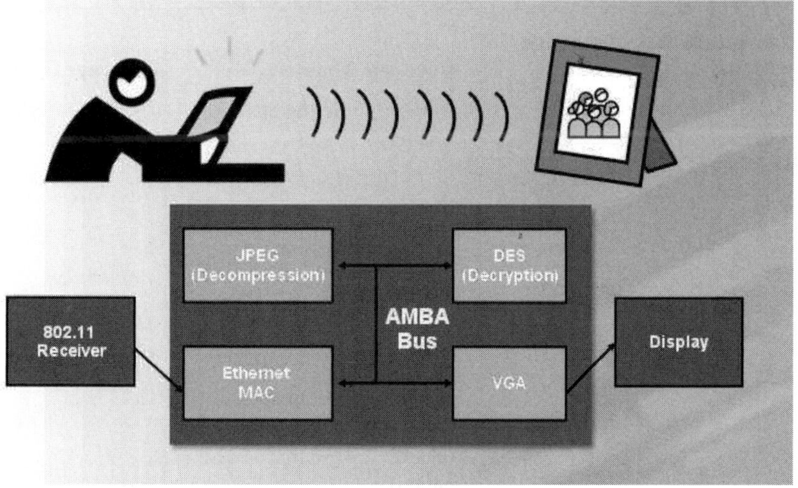

Figure 8-1. Block diagram

The first step to in the functional coverage process is creating a list of functionality that needs to be exercised. A partial list for this device is given in the table below.

Table 8-1.

Coverage Model	Location in Specification	Description	Type of Functional Coverage
DMA Operations	Line 5b	Prove that all possible DMA operations have occurred.	Specification Item
Ethernet operations	Line 8	Prove that Ethernet traffic has occurred for both transmit and receive ports on both Ethernet ports.	Specification Item

The next step is creating a coverage model for each of the items. The coverage model for the DMA operations is given below.

Table 8-2.

Coverage Model	Coverage Items	Coverage Monitors	Coverage Goals
DMA Operations	dma_source	Ethernet to SP (Scratch Pad Memory)	1
	dma_target	SP to DES (Decryption block)	1
	dma_count	DES to JPEG Memory	1
	dma_done	JPEG Memory to VGA memory	1
		JPEG Memory to Ethernet	1
		Generic DMA Operation (i.e. a DMA operation regardless of source or destination)	15

8.3.4 The role of Functional Coverage and Code coverage data in the verification environment

The difference between Functional and Code coverage has been highlighted in section 8.2.1. Let's examine how they can be used together to gauge the completeness of a verification effort.

- The two are complementary:
    - 100% function coverage — does not imply 100% code coverage
    - 100% code coverage — but still not achieve functional-coverage goals

Functional Coverage	Code Coverage	Interpretation
Low	Low	Need to run lots more testcases
Low	High	Need to run more functional testcases
High	Low	Need to enhance quality of verification plan
High	High	Getting pretty close...

Note: If Code and Functional Coverage are both high, but low on assertion coverage, then functional coverage Model definition too vague (need to enhance quality of model)

8.4   Functional Coverage Tools.

8.4.1 Commercial Functional Coverage Tools.

Functional coverage tools give the user the ability to determine what functionality is exercised during verification regardless of the structure it executes. There are very few companies that supply tools in the area. My experience has been that most companies develop ad- hoc strategies for functional coverage. These typically include instrumenting your code to write information to a log file and then creating scripts to extract and analyze the information collected. The commercial tools on the market are split into two camps: collection of coverage data around coverage points and collection of coverage data around transactions.

Coverage points are entities in the design that have meaning for specific functionality. For example if you are exercising a processor an area of functional concern may be the execution of specific sequences of instructions. In this case the instruction register may be selected as a coverage point to observe the particular instructions that are being executed. Once the instruction information is captured it can be analyzed to make sure the appropriate sequences of instructions were executed during verification.

Transaction level functional coverage is applied at a level of abstraction above the design entities. In this form of coverage the intent of the sequence is captured along with the sequence. This allows the user to ask most complex questions about interactions of functionality between different parts of the design.

8.4.2   Features of a good functional coverage tool.

Some of the desired features of such a tool would be to provide the ability to:

- Create and capture a Coverage Model

  The contents of a typical Coverage Model is defined in section 8.3.1. A good functional coverage tool should have the ability to provide the verification person with the capability to capture functional coverage models and explicitly state the conditions for collection of coverage data on the targeted condition and sampling event.

- Specify Coverage goals

  At the start of simulation, coverage goals are specified for each coverage item. The tools should have the capability to specify the coverage goals for each coverage item and record it as a numerical value, or range of values as specified in the verification plan.During simulation all coverage data collected is compared to the target to determine if any progress is being made in the current simulation run.

  Say for example, if after N number of simulation cycles, the targeted coverage on any item has not increased by the desired amount, then we are spinning our wheels. For this purpose, incremental-coverage goal is also an important entity that helps gauge the success rate. Failure to progress (FTP) with the current random stimulus, or random number seed can be determined say if after a specified amount of simulation cycles, the coverage value has not gone up more than the desired value.

- Identify Coverage holes

  One of the most important aspects of functional verification is to identify holes in the coverage space. The inability to reach targeted coverage goals is interpreted as coverage holes, which can be directly correlated to missing, or inadequate stimulus. The goal for any successful verification effort is to achieve the specified target goals with the least amount of simulation cycles.

  If the coverage items target architectural entities, say for example a certain kind of packets on an interface, or a certain transition in a state machine, then this can help identify coverage holes in the stimulus space that fail to excite the circuitry in the desired fashion. This is a lot better than raw code coverage data. For example, simple code coverage can determine if the packets have arrived on the above interface, and that RTL code has toggled to create the desired transition in the state machine, it can not interpret the correctness as can be done by functional coverage above.

- Ability to specify search criteria

  The ability to perform complex parameterized searches on the collected simulation data is very important. Queries are typically performed based on a clearly stated search criteria. The tool should have the capability to capture search criteria in code, and then perform the search on multiple simulation databases.

  In the Incisive environment, the Transaction Explorer (TxE) tool can perform complex searches on the transactions created and stored during different simulation runs. For example, DMA transactions for a certain SRC/DST pair can be queried for, which are then used to update corresponding coverage numbers.

- Ability to do cross-product and binning

  The tool should have the capability to create cross-products between individual coverage items - what does cross-product tell us? In cases where coverage is collected over a wide range of parameters, for example range of memory addresses, binning is a useful concept that helps quantize what is being covered to specified ranges. An example would be to create bins for access to memory address ranges - let's say for example spanning addresses 0x0000 to 0xF000. Individual coverage on each memory address is not only cumbersome but meaningless, and does not convey any important information. However, if one were to create bins spanning sub-ranges, say for example 0x0100 to 0x01F0 corresponding to control registers, and 0x1000 to 0x5FFF corresponding to external memory, then all of a sudden this concept of binning becomes meaningful. So, during simulation, if we had N number of hits within the address range of addresses corresponding to control regs, it can potentially be interpreted as test stimulus having caused N number of accesses to the control registers.

Now, if one were to take it one step further, say by create a cross product of control-reg accesses with say DMA activity on the bus, or with a certain state machine being in a certain state, this conveys meaningful information about the relationship between DMA activity, or state machine state with control-register access.

So, how can this information be used by the verification engineer? Let's look now at the verification plan. If the target for coverage for control-register access is set to a value X, then by observing cross coverage information, the verification engineer can use this information to constrain the generation of random stimulus, say on the bus to the specified range that would get him the targeted coverage X on the control register access. On the other hand, if one relied on purely random generation of stimulus, without any constraints on the generation of stimulus, the chance of generation of X hits on the control-registers is non deterministic, and can take many simulation cycles to achieve without any guarantees of generating X number of desired hits to the control-regs.

8.4.3 Requirements for a Functional Coverage Tools

A functional coverage tool must have a certain set of requirements to be effective. This minimum set and their reasons are listed as follows:

- Easy capture of coverage data

  Coverage data needs to be captured before it can be used to create a metric. The process of capturing the data should be straight forward. The tool should have enough flexibility to capture arbitrary data, for example test information, the state of the entire system, as well as signal information.

- Easy creation of coverage metrics

  A coverage metric needs to be created from the data captured. The mechanism for this should be understandable. Creation of these searches, if you will, should be able to be done very quickly, measured in minutes. The searches should be reusable across designs with the same or similar interfaces and features.

- Creation of coverage metrics should be reusable across multiple designs of similar types

- Easy analysis of coverage data

  While creating a functional coverage metric is the overall goal of this type of tool it is also useful to have additional analysis capabilities. For example it would be useful to graph some of the results. If a particular test does not achieve the coverage expected it would be useful to use to apply the same metric interactively to the waveforms to help determine why the test did not achieve the desired result.

- Ability to accumulate results across multiple simulation runs

  Functional coverage metrics show how much of the functionality was exercised by either a single simulation run or the entire regression suite. Therefore any tool that creates functional coverage metrics needs to allow results to be accumulated across any number of simulation runs.

- Creation of reports

  The basic report that needs to be created is a summary of all of the functional coverage elements. Additional reports help to understand what is being presented.

- GUI

  A GUI is useful for the novice that is accessing the functional coverage tool for the first time and sometimes useful for advanced users who are creating complex searches for functional behavior. The GUI should help the user create a simple functional coverage search. It should also help the user create various functional coverage reports.

- Ability to access the functional coverage tool during simulation and in a post-processing fashion.

  To achieve the most results from a functional coverage tool it should be able to operate during simulation, allowing access to its results by the testbench, as well as perform analysis after simulation. Coverage information is useful during simulation for two reasons. First it could be used by the test to determine when to stop. The test can stop when the coverage hits a certain level. Second it can potentially be used to alter the test, effectively allowing the test to react to the coverage information. This is only viable if there is an easy correlation between a coverage item and parts of the test. For example, take the case where the coverage a certain of number of ethernet packets to go from input port 0 to output port 1. If this coverage goal is met then the constraints on the randomly generated addresses could be changed such that the address range for output port 1 is never again selected. Now lets take the case of an arbitration circuit buried deep inside the design. If it is difficult to control the arbitration from the primary input then the test could not react to the coverage information.

8.5   Limitations of Functional Coverage

There are several limitations of functional coverage that should be taken into account. Some of them are:

- Knowledge of what functionality to check for is limited by the specification and the knowledge of the design.
- There is not a defined list of what 100% of the functionality of the design is and therefore there may be functionality that is missing from the list.
- It may take too long to create a coverage model to justify the investment.
- There is no real way to verify that a coverage model is correct except to manually check it.
- The coverage goals are not always easy to decide upon if the goal needs to be greater than one.
- The effectiveness of functional coverage is limited to the knowledge of the implementor.

CHAPTER 9

Dynamic Memory Modeling

One of the performance challenges in functional simulation is simulating large memories in the design that take too much simulation memory space. This chapter focuses on providing a solution for using memory models for optimum simulation performance. These memory models can be either bought from memory solution providers or they can be built in house by designers. These are mainly required for:

- Doing architecture analysis.
- To assist in selecting memory architecture for the design.
- Regression running simulations for verifications.

Following topics are covered in this chapter:

- Various alternatives for using memory models in simulation.
- Running simulations with memory models.
- Building dynamic memory models inhouse.
- Buying commercially available solutions.
- Advantages/disadvantages of different solutions available.
- Dynamic memory modeling techniques.
- Examples of dynamic memory models using Testbuilder.

9.1   Various solutions for simulation memory models

All system designs contain embedded or standalone memories. Since, it is part of almost every chip that is manufactured, industry demands more proficient solution for modelling memories and unifying them. There are several ways, discussed as follows, for making these memories available to design teams

- Use models provided by vendors: ASIC vendors will often provide HDL models for embedded memories as a part of the design-kit. In addition, some memory vendors, including Micron and Intel, will often provide HDL models available for free download from their web sites. These models will represent the memory components manufactured by these vendors and will often cover some fraction of the components manufactured by the vendor

- Build memory models in house: Designers can choose to build their own memory models from scratch or acquire a free one and build a wrapper around it for accuracy and better performance.

- Buy accurate memory infrastructure from outside: There are companies that often provide highly accurate models for memory for various memory technologies and components that are available. They have established relationships with memory vendors to be acquainted of any additions/modification to memory specifications, even before they are released. One of the leading companies for memory solutions is Denali.

9.2   Running simulation with memory models

For running simulation with memory models, either built in models can be used or these can be bought from outside. Following sections discuss about both these cases.

9.2.1   Built in memory models

To run simulation with memory models from vendors, you must have simulation licenses from them. Simulation models are necessary for system designers who first model various elements in the system in software and then run simulations using these to certify correctness of the system before silicon production. For most of the designs at this architectural stage, memories are not so complex and can be modeled easily. Also, mostly designs do not have very complex memories to be modeled. Knowledge of how to build dynamic memory model in house can be used to model accurate memory models with better performance. Either of following methods are used for running simulations at this point:

9.2.1.1 Use verilog models provided by vendors

These are HDL models provided by ASIC vendors or memory vendors. These verilog/VHDL models can be easily used in simulation environment. But simulation becomes slower because of large memory space from memory declarations of these models. Also, they usually have a large number of protocol checks inside to make sure that the accesses to the memory are correct which slows down the performance too.

9.2.1.2 Wrapper for performance

Designers have choice of using above models and use their own resources to fill in the gaps as needed. Mostly these gaps are the degradation of simulation performance. This can be overridden by writing a wrapper around verilog models to convert it to dynamic memory. This way, it is possible to get accuracy of verilog models provided and also achieve the performance. Example below shows how a small wrapper is written in Testbuilder tool to convert any verilog memory model to dynamic form. It shows amazing results in terms of performance and memory utilization as compared to using original verilog model or using the one provided by memory solutions provider. It is better than original memory model because of dynamic declaration of memory and it is better in performance from the bought memory model since there is no overhead of tool in supplying lot of other features provided.

9.2.1.3 Using C models

Above solutions do not take much extra time once methodology for integration is there. If designers choose to write their own models from scratch, it can take a lot of time and effort to put down the whole methodology for developing and testing/ debugging model in house. The challenge is to not only write a very appropriate C code that gives very good performance and memory utilization but guarantee completeness in terms of covering 100% data sheet for both timing and functional information.

Below are listed some of the main challenges for developing highly accurate model:

- It is required to understand the functional behavior of the device very well.
- All the fine nuances in the datasheet should be well understood and covered as part of the check.

- It should be seamless to make any change in memory model is the spec changes.

- Should have appropriate testcases for verification of these models.

- Some time has to be spent on debugging and fixing the models.

Based on the current phase of the design, it might not be essential to have a highly accurate model that throws exceptions thoroughly after power up and checks states for all the control signals. But it is essential that the designer who is modeling memory has an expertise to understand and differentiate all aspects of data sheets from various vendors.

As an example for DDR there is standard JEDEC specifications that is out there, but in spite of this every vendor has their own road map that they support which deviates from JEDEC specifications. The DRAM market is extremely fragmented and taking all these factors into considerations, it becomes extremely challenging to have quality models which can understand these functional details amongst different vendors.

For example, every vendor has a different definition in terms of how they support auto-precharge functionality when it comes to write with auto-precharge or read with auto-precharge being interrupted by a write or read sequence. Different vendors have different performance benefits with this functionality. Behavior of this capability also differs within the same vendor as well, depending on the memory density, data width, frequency of operation etc. With some vendors and certain parts it can be interrupted in the middle of the burst, with certain parts it can be interrupted at the end of a burst or certain parts it has to wait till the write recovery time is over say after write with A/P sequence. Even more confusing and complicated is the fact that certain vendors even support asynchronous auto-precharge versus synchronous precharge. This means that in the asynchronous case the precharge begins internally much before and extra cycle can be squeezed in giving a better performance. Modelling all this is not easy in terms of reaping the best performance and requires ability to understand these differences and how to incorporate these in the model.

This is just one scenario, there are lots of other differences like supporting traslockout or not in which case some parts have internal circuitry to delay the internal precharge during a read operation if tras is not satisfied by then or delay the internal precharge till it is satisfied. Some parts do not have this functionality, so these do not have this level of flexibility. For thorough DDR initialization checks, it is required to check for the state of every control signal from powerup itself till initialization sequence is complete. This is very crucial factor because if the init sequence which itself is very robust is not in proper order and not checked for accordingly then the device will already be in undefined state and then no matter how much time and effort is put to check for correct device behavior for the rest of the simulation, it will not help because device is already malfunctioning from the start itself.

9.3   Buying commercially available solutions

If designers do not want to build these models themselves and instead invest their time in design efforts, these can be bought from the vendors. These vendors work in association with semiconductor manufacturing companies to provide higher quality models that represent the memory behavior more accurately. Accuracy is more important in designs for applications like set-top boxes, servers, workstations and wireless need highly accurate simulation models of flash memories used in the system. Mostly these companies are working with memory vendors way before the memory is released. So that in late design cycle you have the model available.

Once it is decided to invest in buying memory model for simulations, one should look for following:

- These should be compatible with various simulators and workstations in the industry.
- Should use dynamic memory allocation for faster simulations.
- Should have consistent behavior in any simulation environment.
- Good to have proper user interface and ease of use features.
- Intelligent automatic error checking and debug capabilities.
- Ensure accuracy of device timing and functionality.
- Should be designed to a proven standard and must have gone through rigorous testing.
- Effortless migration to new memory architectures.
- Support should be available for broad range of memories
    - Flash memories
    - SRAMs
    - FIFOs
    - Synchronous DRAM
    - Synchronous graphics RAM
    - Rambus DRAM
    - Serial and parallel EEPROMs

- Content-Addressable Memory

- Multiport synchronous SRAM(quad-port, dual-port)

- RLDRAM

Accessing these should be easier for the designers. Denali provides such support through ememory.com. Denali models also integrate with SCV, testbuilder and synopsys Vera environment and other available verification tools.

9.4   Comparing Built in and Commercial memory models

Based on the requirements, resources and the schedule, the memory model is either built in house or bought from the memory vendor. Built in is mostly used for architectural analysis and not for detailed accurate simulation runs. The flow for built in is as follows in figure 9-1

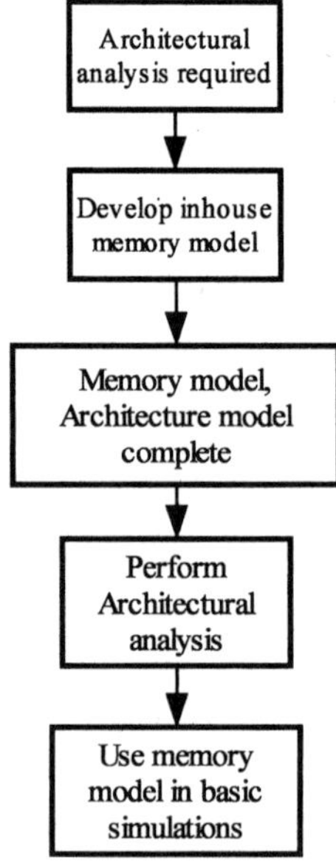

Figure 9-1. Built in memory models flow

Advantages of built in:

- Seamless integration of built in models to already used simulation environment.
- There is no impact on existing environment to integrate external component.
- Don't have to wait for availability of support.
- Better performance since have control over the debugging.
- Faster simulations and integration.
- Regression running is not limited by number of memory model licenses available. Can run hundreds of regression jobs without extra cost of licenses.

Disadvantages:

- Cannot guarantee functionality and accuracy of these simulation models.
- Resources have to be spent on modeling and testing these.

Commercially Available Memory model: These are very accurate and easy to download and integrate. Following flow in figure 9-2 represents the model usage from Denali:

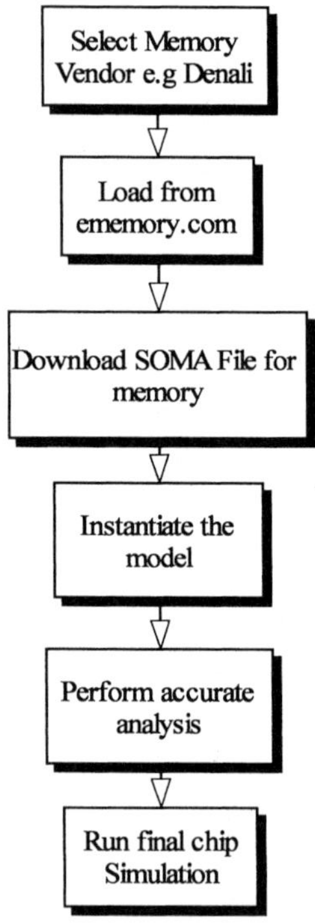

Figure 9-2. Memory model flow from Denali

Advantages of outside:

- Proven, used before and tested.
- Have relation with memory vendors to nail down the detailed issues for modeling.
- Cover all aspects of data sheet, understand their road map better in terms of which parts are going to be phased out, getting data sheets for next generation memories etc.
- Can share technical notes regarding specific implementation of various memory parts.

- Not only regress vendor models inhouse but also provide vendors with copy of model software that can be run against their regression suite.
- Any issues found in above regression are fast tracked through bug fix programs.
- Past usage of models build more confidence in them since they have been tested.
- Work on the memory with vendor much before it is released. At the time data sheets are made public, models are already available.
- Timely notification to customers about parts phasing out to give them ample time to switch to another compatible part with minimal impact to the design.
- Full protocol checking for power down and self refresh modes.
- 100% coverage of data sheet timing checks.
- Dynamic simulation cycle time modeling.

Disadvantages:

- Regression needs hundred of jobs running simultaneously. Need to buy hundreds of licenses for simulations
- Spent time and effort in integrating new tool environment
- Time and effort to learn using new tool environment

Wrapper Memory model: Incase verilog or VHDL memory model is available from Vendor, it can be used as it is or converted to the dynamic memory model by wrapping it up with C/C++ memory declarations. Flow is shown in figure 9-3.

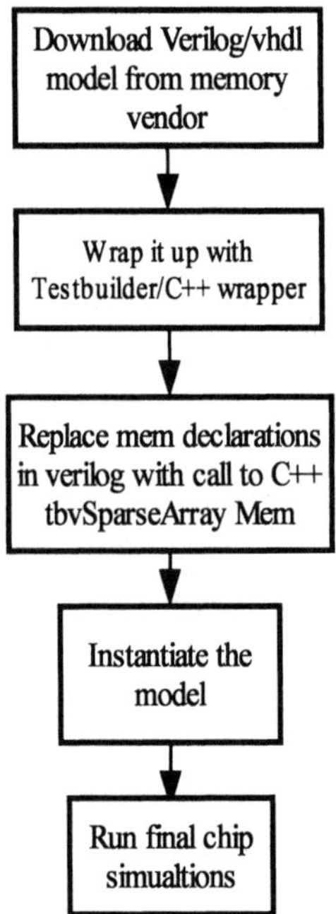

Figure 9-3. Wrapper memory model flow

Advantages of wrapper:

- All the advantages for built in model.
- Getting the accuracy of memory vendor verilog model.
- Better performance.
- No extra costs.
- Almost no extra time and effort once wrapper methodology is in place.

Disadvantages

- Dependency on accuracy of model provide by vendor who manufactured the memory or where the model is downloaded.

9.5   Dynamic Memory modeling Techniques

9.5.1 Verilog

Large models require long simulation time and large amount of system memory to simulate.

For optimum simulation performance, memory can be programmed so that only the active part of memory is stored in the data structure:

- Using PLI.
- Using the $damem dynamic memory task in verilog-XL.
- Using mapping schemes with standard verilog HDL to maintain only a subsection of the entire memory.

Verilog-XL comes with PLI called "damem". Here are some of the excerpts from the appnote.

System tasks:
$damem_declare
$damem_read
$damem_write
$damem_initb
$damem_inith
================================
The $damem_declare System Task
================================
The $damem_declare system task declares a memory in your design. It is an alternative to the use of the reg keyword to declare an array of regs. When you use the $damem_declare system task to declare a memory in a design, Verilog-XL does not allocate memory space in your workstation for that memory in your design until you enter another system task for dynamically allocated memories. Here is the format:

$damem_declare(name_of_memory,bit_from,bit_to, addr_from, addr_to);

$damem_write: This writes data to the memory declared by above system task. Here is the format.

$damem_write (name_of_memory,addr,value);

$damem_read: This reads data from memory declared by damem_declare system task. Here is the format:

$damem_read(name_of_memory,addr,register);

$damem_initb or $damem_inith: This initializes memory to binary or hex value.that writes data to that memory in your design.

While allocating above memory, Verilog-XL only allocates enough memory space to hold the data written to that memory in your design.

9.5.2 Using Wrapper

Example of memory modeling using Testbuilder wrapper for verilog:

The tbvSparseArrayT data structure lets you model extremely large memory elements efficiently. SCV has equivalent scv_sparse_array to be used. The data that is stored in the array can be as simple as an integer value or as complex as a user-defined class representing a packet in a communications protocol.

Sparse arrays are indexed using 64-bit unsigned long values. For arrays that need larger address spaces, you can use an associative array to mimic the functionality (see Associative Arrays).

```
tbvSparseArrayT<int> sa("my_array",0);
sa[454] = 10;
    sa[458] = sa[454] + 100;
    // Right now, memory is allocated for two elements:
    // sa[454]=10 and sa[458]==110; sa[others]==0.
    sa.erase(454);
    // Now memory is allocated for one element:
    // sa[458]==110 and sa[others]==0.
    for (tbvSparseArrayT<int>::iteratorT iter = sa.begin(); iter != sa.end();
++iter) {
        int i = iter->first;    // Get the index of the element.
    int e = iter->second;   // Get the element.
    }
    sa.clear();
```

This example shows a common memory example of an array of two-state signals.

Step 1: To define a sparse array :

// Create a sparse array of four-state signals with a default

// value of x, 8 bits wide.

```
tbvSparseArrayT<tbvSignal4StateT> sigArray("SignalArray"
tbvSignal4StateT(7,0,"0xxx"));
```

Step 2: To populate the memory array.

This example fills in the addresses 512-2047

```
// Starting with address 512, add signals up to, but not including,
// address 2048, filling in every 32nd location.
for(int i=512; i<2048; i+=32) sigArray[i] = (i >> 1) & 0x000000ff;
, but it only fills in every 32nd address.
```

Step 3: is letting the HDL access the memory.

testbench must let the simulation know that there is information in the memory, so that the simulation will access that information

```
// The HDL uses the "getVal" event to get a value out of the
// array. The HDL assigns the "addr" signal to the address it wants
// to get and expects the data to be placed on the bus.
tbvSignalHdlT addr("top.addr"),
bus("top.bus"),
done("top.done");
tbvEventT getVal("top.getVal");
tbvWaitAnyT w;
w.add(done);
w.add(getVal);
while(done == 0) { w.waitEvent(); if(w.which()==1) bus=sigAr-
ray[*(addr.getValueP())]; }
```

Step 4: is communicating with memory model. the following is the Verilog code that communicates with the sparse array memory model in the Testbuilder C++ testbench.

```
module top;
// Add some Verilog signals to place in the array.
reg [7:0] bus;
reg [7:0] addr;
event getVal;
initial begin done=0;
```

```
$tbv_main; end
// Change the bus from the C++ and look at the output here. The bus
// takes on the values of the sparse array. Location 2048 should be x.
initial begin
#10;
for(addr=1024; addr<=2048; addr=addr+128)
begin
->getVal;
#1 $display("addr[%d] = %x", addr, bus);
end
done=1;
end
endmodule
```

9.6   Example From Reference Design:

The example below uses the dynamic memory model that was developed by wrapping up the existing downloaded verilog model from Samsung. This model was quite accurate for the simulations required and wrapping it in Testbuilder code increased the performance for simulations much more than verilog model or commercial available memory model. All the external memories in the reference design were replaced with similar wrapper.

For running regression

- Specify what to run with, options of internal memory, commercial memory model or built in models.

- Select the tests that require models from commercial memory vendor and give command line option.

- For rest of them use built in models.

Description of the files:

1. memModel.cc:

2. memModel.v

3. memModelTest1.cc

4. k7n163645a_R03.v: Samsung verilog model provided for the SSRAM part number K7N163645A

5. k7n163645a_test_R03.v: testbench provided with above model

Testbuilder command for running this simulation is:

testBuilder.sh memModel.cc memModel.v memModelTest1.cc
k7n163645a_R03.v k7n163645a_test_R03.v -ncv "+tcl+ncrun.scr +define+hc20"

The verilog HDL model for SSRAM is downloaded and is wrapped using the Test-builder dynamic memory. Here is the file memModel.v used for wrapping it up:-

```verilog
'timescale 1 ns / 1 ns
module memModelTvm(r_wb,
          addr,
          d,
          q,
          trigger_rd,
          trigger_wr,
          data_bits,
          mem_size_high
          );
input        r_wb;
input [18:0]   addr;
input [35:0]   d;
output[35:0]   q;
input        trigger_rd;
input        trigger_wr;
input [35:0]   data_bits;
input [18:0]   mem_size_high;
reg        doExit;
reg [35:0]   q_r;
assign q=q_r;
  //
  //  register every instance of this TVM to test builder
  //
  initial begin
      $tbv_tvm_connect; // connects C++ defined TVM
      doExit = 0;
  end
endmodule
```

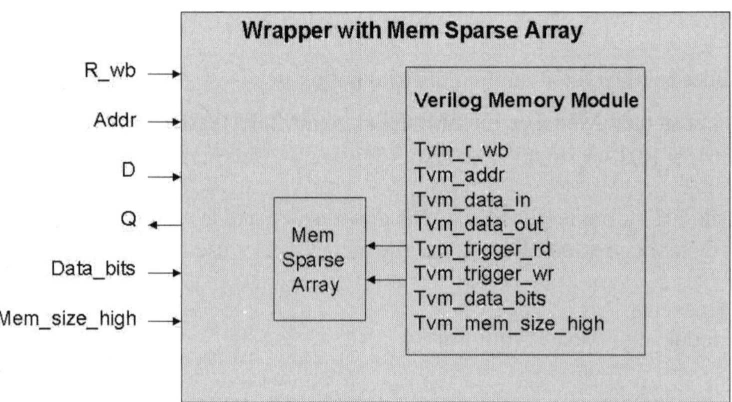

Above TVM is used to wrap up the Verilog/VHDL model downloaded. This TVM talks to the C++ defined dynamic memory allocation. In the verilog module k7n163645a_R03.v declaration for memory is replaced by call to this wrapper.

reg   [`data_bits -1 : 0] mem [0 : mem_sizes]; This declaration, from original memory model , is used no more.

Following lines are added at the top verilog module:-

```
// Declare the signals for connecting to wrapper
wire    [(`data_bits - 1) : 0] tvm_data_out;

reg    [(`data_bits - 1) : 0] tvm_data_in;
reg[(`addr_bits - 1) : 0] tvm_addr;
reg tvm_r_wb;
reg tvm_trigger_rd;
reg    tvm_trigger_wr;

reg[(`data_bits - 1) : 0] tvm_data_bits;
reg[(`addr_bits - 1) : 0] tvm_mem_size_high;
// Initializing for different inputs
initial
begin
tvm_data_bits = `data_bits;
tvm_mem_size_high =  mem_sizes;
tvm_trigger_rd = 0;
            tvm_trigger_wr = 0;
```

```
end

    // instantiate the memModel TVM for dynamic memory allocation
    memModelTvm memModelTvm0(.r_wb(tvm_r_wb),
      .addr(tvm_addr),
     .d(tvm_data_in),
     .q(tvm_data_out),
     .trigger_rd(tvm_trigger_rd),
     .trigger_wr(tvm_trigger_wr),
     .data_bits(tvm_data_bits),
     .mem_size_high(tvm_mem_size_high));
```

After initializing, the calls for writing and reading in original verilog module are replaced accordingly.

Modifications in read and write logic is required where assignment for data is done to memory location.

```
//tmp_data = mem[{wburst_add,bcnt}] ; // This is replaced with following code
tvm_addr = {wburst_add,bcnt};
tvm_r_wb = 1;
tvm_trigger_rd=~tvm_trigger_rd;

// mem[{wburst_add,bcnt}] = tmp_data ; This is replaced with following code
tvm_r_wb = 0;
tvm_data_in = tmp_data;
tvm_trigger_wr=~tvm_trigger_wr;
```

Here is the testbuilder code in C++ for memModel.cc file:-

```
#include "memModel.h"
#include "TestBuilder.h"


// ****************************************************************
//                memModelTask
// ****************************************************************
memModelTaskT::memModelTaskT ( memModelTvmT & tvm ) : tbvTaskT(),

    // TVM reference
    memModelTvm ( tvm )
{
    // Disable the mutex built into the fiber and turn off automatic
    // transaction recording
```

```cpp
  setSynchronization(FALSE, FALSE);
}

memModelTaskT::~memModelTaskT () {}
void memModelTaskT::body(tbvSmartDataT * dataP = NULL)
{
  // Do one transaction at a time
  memModelTvm.memModelMutex.lock();

// Define events for triggereing read and write on any edge of trigger_wr or
trigger_rd signals from verilog.

                  tbvEventExprT    trigger_R    (memModelTvm.trigger_rd,
tbvThreadT::ANYEDGE);

                  tbvEventExprT    trigger_W    (memModelTvm.trigger_wr,
tbvThreadT::ANYEDGE);

    tbvEventExprT trigger (trigger_R || trigger_W);

    tbvEventExprT finishedSem (memModelTvm.isRunningSem);
    string z("0b");
    for(unsigned int i = 0;i<=(*(memModelTvm.data_bits.getValueP())-1);++i){
      z.append("Z"); // initialize to "Z"
    }
    tbvSignal4StateT  zed((*(memModelTvm.data_bits).getValueP())-1,0);
    zed = 0;
// Define memory bank below using sparsearray to dynamically allocate mem-
ory

mem_bank          =          new          tbvSparseArrayT<tbvSignal4StateT>
("mem_bank",zed,0,*(memModelTvm.mem_size_high.getValueP()));

// For simultaneous threads of different memory model TVMs running in simu-
lation

    while(memModelTvm.isRunning()){
      tbvEventExprT finished (finishedSem || trigger);
      tbvEventResultT r = finished.wait(tbvEventExprT::RWSYNC);
      r.release();
// At trigger do read or write in the array
      if(r.getActive(trigger)){
      if(*(memModelTvm.r_wb.getValueP())){
      memModelTvm.q = (*mem_bank)[*(memModelTvm.addr.getValueP())];
```

```cpp
      }else{
              (*mem_bank)[*(memModelTvm.addr.getValueP())] = memMod-
elTvm.d;


      }
    }
  tbvThreadT::yield();
  }
  // release the semaphore
  memModelTvm.memModelMutex.unlock();
}
// *****************************************************************
//                  memModel TVM
// *****************************************************************
//
// TVM memModel Constructor & Destructor
//
memModelTvmT::memModelTvmT ( ) : tbvTvmT(),

  // Initialize the TVM Signals from verilog
      r_wb (getFullInterfaceHdlNameP("r_wb") ),
      addr (getFullInterfaceHdlNameP("addr") ),
      d (getFullInterfaceHdlNameP("d") ),
      q (getFullInterfaceHdlNameP("q_r") ),
      trigger_rd (getFullInterfaceHdlNameP("trigger_rd") ),
      trigger_wr (getFullInterfaceHdlNameP("trigger_wr") ),
      doExit (getFullInterfaceHdlNameP("doExit") ),
      data_bits(getFullInterfaceHdlNameP("data_bits") ),
      mem_size_high(getFullInterfaceHdlNameP("mem_size_high") ),
  // Initialize the Mutex
  memModelMutex( "memModelMutex" ),
  // Create the task objects
  memModel(* this),
  running(false),
  isRunningSem("isRunningSem",1)
{
}
//destructor for tvm memModelTvmT
memModelTvmT::~memModelTvmT ()
{
  //tbvThreadT::waitDescendants();
  //add any clean up code needed for tvm linkTMon
}
```

```cpp
// Call these to start or stop the mem model tvms any time during simulation
void memModelTvmT::start(){
     running = true;
     isRunningSem.wait();
}
void memModelTvmT::stop(){
     running = false;
     isRunningSem.post();
}
bool memModelTvmT::isRunning(){
     return running;
}
// Functions for calling read and write to memory
bool memModelTvmT::write(unsigned int address,tbvSignal4StateT & data){
     (*(memModel.mem_bank))[address] = data;
     if((*memModel.mem_bank)[address] == data)
       return true;
     else
       return false;
}
tbvSignal4StateT & memModelTvmT::read(unsigned int address) {
     return (*memModel.mem_bank)[address];
}
void memModelTvmT::readArray(){
     tbvOut<<"array content"<<endl;
      tbvSparseArrayT<tbvSignal4StateT>::iteratorT iter;
  for (iter = (memModel.mem_bank)->begin();iter != (memModel.mem_bank)-
   >end();++iter)

  {
           tbvOut<<"mem_bank["<<iter->first<<"] = "<<iter->second<<endl;
       }
}
//
// Method called from $tbv_tvm_connect, to create the TVM object.
//
void memModelTvmT::create ( )
{
  new memModelTvmT ( );
};
```

memModel.h file:

```cpp
#ifndef MEM_MODEL_H
#define MEM_MODEL_H
```

```cpp
#include "TestBuilder.h"

class memModelTvmT;
// ***************************************
// Task memModelTaskT
// ***************************************
class memModelTaskT : public tbvTaskT{
  public:
    //parent tvm, make it easy to access in task body
    memModelTvmT& memModelTvm;

    //constructor, destructor
    memModelTaskT ( memModelTvmT& );
     virtual ~memModelTaskT ();
     tbvSparseArrayT<tbvSignal4StateT> * mem_bank;

     //body function for task functionality
     virtual void body (tbvSmartDataT * dataP = NULL);
};


// ***************************************
// TVM memModelTvmT
// ***************************************

class memModelTvmT  : public tbvTvmT
{
  public:
    //constructor, destructor
    memModelTvmT ();
    virtual ~ memModelTvmT ();

    //HDL interface signals
tbvSignalHdlT   r_wb;
tbvSignalHdlT   addr;
tbvSignalHdlT   d;
tbvSignalHdlT   q;
tbvSignalHdlT   trigger_rd;
tbvSignalHdlT   trigger_wr;
tbvSignalHdlT   doExit;
tbvSignalHdlT   data_bits;
tbvSignalHdlT   mem_size_high;

    // The Mem Model Mutex
    tbvMutexT memModelMutex;
```

```cpp
void            start();
void            stop();
bool            isRunning();
bool            write(unsigned int,tbvSignal4StateT &);
tbvSignal4StateT&   read(unsigned int);
void            readArray();
bool            running;

// the recording fiber is needed to avoid using the parent task fiber
tbvFiberT memModelFiber;

memModelTaskT   memModel;
tbvSemT         isRunningSem;

// This function is called from $tbv_tvm_connect, to Create a TVM object.
 static void create ( );

};
```

Following code in testbuilder is for simulation that calls the c++ memory allocation wrapper and verilog code for memory model and testing:-

```cpp
#include "TestBuilder.h"
#include "memModel.h"
void tbvMain()
{
  // begin test
  tbvOut << "C++: Starting test at t="
      << tbvGetInt64Time(TBV_NS) << " ns" << endl;
  // access the TVM through the test builder registry
  memModelTvmT& memModelTvm0 = (memModelTvmT&)
*tbvTvmT::getTvmByInstanceNameP("memModelTvm0");
  memModelTvm0.memModel.spawn();
memModelTvm0.doExit = 1;
  tbvThreadT::waitDescendants();
  tbvWait(100);
  tbvExit();
};
//
// The list of TVMs to be instantiated from Verilog.
```

```
//
tbvTvmTypeT tbvTvmTypes[] = {
 { "memModelTvm", memModelTvmT::create, 0 },
 { 0, 0, 0}
};
// The tbvInit() is called before anything else. This is useful
// for initialization
void tbvInit(){}
```

Testbuilder has integration with Denali data driven interface. Denali verification functions are directly accessible from Testbuilder and interface, this also supports callback capabilities such that when ever any transactions like read/write/error injection is happening at the memory interface, a notification is issued within Testbuilder environment based on which directive testing can be done and performance metrics can be run. A lot of users who use Denali with testbench tools use this callback for scoreboarding. An expected list of ordered transactions can be setup through Testbuilder and everytime there is a callback with the address location, data/mask/unknown bits for a type of access like read/write/error injection, it can be checked against an expected list.

9.6.1 Performance comparison

Performance comparison for different memory models was done for speeding up the regression performance and to be able to run more regression jobs simultaneously than given commercial memory model licenses available at that time. For above example, simulations were done for one basic testcase using all different memory models. This performance comparison evaluates memory utilization and performance.

Figure 9-4. Resource usage For Commercial/wrapper Verification memory models

	Commercially available Memory Model	Memory models using TestBuilder	Verilog Models
CPU Time	1200.20 sec	1151.87 sec	1528.7 sec
Max Memory	257 MB	250 MB	720 MB
Max Swap	285 MB	305 MB	751 MB

As seen, for one simple testcase that runs for 20 minutes verilog model take lot of swap space. Normally most of the tools have limitation of not being able to support more than 4G of job. With this kind of swap space utilization, for long test in regression this limit will be easily reached.

CHAPTER 10

Post Synthesis Gate Simulation

First necessary step in design verification is functional simulation. Once functional requirements for design are met, timing verification is essential to determine if design meets timing requirements and better design performance can be achieved by using more optimized delays. This chapter focuses on the dynamic timing simulation technique that can verify both the functionality and the timing for the design. It involves exhaustive simulation on back annotated gates.

Two timing verification procedures are mostly used - the static timing analysis and the back annotated gate level simultions. Where static timing analysis (STA) is used to verify timing of the entire device and checks every path in a design or any timing violations without checking for the functionality. This uses the sophisticated analysis features such as false path detection and elimination and minimum/maximum analysis. STA can verify timing requirement for only synchronous circuits. Once confidence is gained on design timing using STA and formal verification techniques, still there are times when it is needed to run gate level simulations especially for generating test vectors for ATE testing. However, there are some challenges associated with running gate level simulations at system level

- Gate simulations are slow.
- Completeness of simulation can not be guaranteed.
- Poor coverage.

Following topics are covered in this chapter:

- Need for gate level simulation.
- Different models for running gate level simulation.
- Various types of simulations.
- Environment setup for gate simulations.
- ATE Vector generation.

10.1   Introduction

10.1.1   Need for Gate Level Simulation

Advances in formal based methodologies have enabled designers and verification engineers to disestablish any need for long gate level simulation to some extent. There is still a need felt for running the gate simulations as specified in the following cases:

- Several assumptions that are made during the design and verification process are not verified by equivalence checking, which makes the use of gate-level simulation essential.

- Certain classes of bugs do not become apparent until the gate level. These bugs include differences in handling of don't-cares or un-knowns between register-transfer-level (RTL) and gate-level designs, as well as functional changes made at the gate level, like test logic and clock trees.

- Even synthesis tools sometimes make some assumptions about the designer's intent when implementing RTL designs into gates. Historically, equivalence checkers can fail to check these assumptions. Generally there might have been RTL don't-cares or unknowns warnings flagged, but when there are too many warnings, the designers often ignore them.

- It is also seen that sometimes the test logic is not added to the design until the gate level, and it may not have a functional counterpart in the RTL design. To equivalence-check the gate- level design against the RTL, test logic is disabled and not verified. Users of equivalence checking must either assume that the test logic is correct, or verify it by running gate-level simulation.

- Waveform database can be saved during full time gate simulation runs at different clock frequencies and can be used later in debugging related to any timing issues seen during ATE testing.

10.2   Different models for running gate netlist simulation

10.2.1   Gate Simulation with Unit Delay timing

The gate level simulation can become extremely time-consuming and resource intensive. So, it is better to start with the unit-delay simulation as soon as the netlist of the design is done even if the chip does not meet the timing requirements. The gate-level simulation with unit delay timing model is generally faster than the full-timing gate simulation but slower than the RTL simulation. There can be various reasons for performing unit-delay gate simulation

- To check if the chip is properly configured and initialized.

- For functional correctness of the netlist and to see that the gate implementation functionally matches the RTL description.

- To prepare the environment for post synthesis simulation. In general it takes a while before the gate level simulation is up and running. This time should not be wasted after the layout is complete.

10.2.2   Gate-level simulation with full timing

The final netlist for the timing simulation is available after the place and route process. Only at this stage of the design implementation can the users see how the design will actually behave in the real circuit. Overall functionality of a design is defined in the early stages but it is not until the design has been physically placed and routed that timing information of the design can be properly calculated with accuracy.

But gate-level simulation with a full timing annotation is very slow. This approach is particularly useful for verifying

- Asynchronous logics.
- Embedded asynchronous RAM interfaces.
- Single-cycle timing exceptions.

It becomes almost unnecessary to run full timing simulations in fully synchronous designs. During full timing gate-level simulation runs, the timing information is back-annotated from the place and route tools and hazards are enabled. Gate simulations are run with the worst-case timing to check for long paths and with the best-case timing to verify any issues related to minimum path delay.

10.3   Different types of Simulation

Simulators are usually categorized for performing one of the following functionality starting from the higher level abstract models to detailed low level transistor-level simulations:

- Behavioral simulation
- Functional simulation
- Static timing analysis
- Gate-level simulation
- Switch-level simulation
- Transistor-level or circuit-level simulation

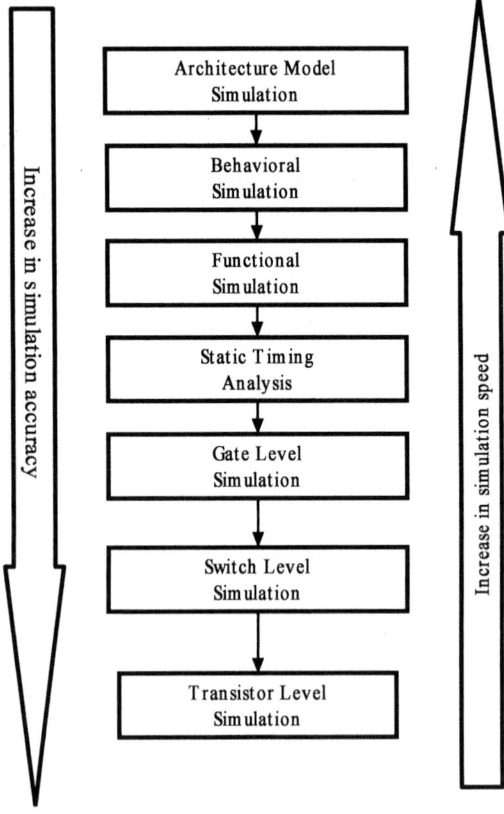

Figure 10-1. Different Simulations

As it goes down the list, simulators becomes more and more accurate but at the same time complexity and the simulation run time increases as shown in figure 10-1. While behavioral level simulation for the entire system can be done at quite fast speed, as you go down the list, it becomes almost impossible to perform the circuit-level simulation of more than a few hundred transistors.

As discussed in previous chapters, there are different ways to create an imaginary simulation model of a system. Functional simulation ignores timing and includes unit-delay simulation, which sets delays to a fixed value (for example, 1 ns). After the behavior is verified for correctness, the system is usually partitioned into sub blocks and a timing simulation is performed for each block separately so that the simulation run times does not become too long. Timing simulation can also be done using the static analyzer that analyzes logic in a static manner, computing the delay times for each path. This is called static timing analysis because it does not require the creation of a set of test or stimulus vectors. But as discussed before, timing analysis works best with synchronous systems whose maximum operating frequency is determined by the longest path delay between successive flip-flops. The path with the longest delay is the critical path.

Timing performance of a design can also be checked using logic simulation or gate-level simulation. In a gate-level simulator a logic gate or logic cell (NAND, NOR, and so on) is treated as a black box modeled by a function whose variables are the input signals. The function also model the delay through the logic cell. Setting all the delays to unit value is the equivalent of functional simulation.

Delay model for the gate decides the accuracy and the speed grade of simulations. Evaluation of elements may be carried out as functional evaluation plus delay evaluation.

In case the timing simulation provided by a black-box model of a logic gate does not provide accurate results, switch-level simulation can be performed which is more detailed level and models the transistors as switches - on or off. Switch-level simulation can provide more accurate timing predictions than gate-level simulation, but without the ability to use logic-cell delays as parameters of the models. The most accurate, but also the most complex and time-consuming, form of simulation is a transistor-level simulation. A transistor-level simulator requires models of transistors, describing their nonlinear voltage and current characteristics.

There are different software tools available for performing each of the above types of simulations. A mixed-mode simulator permits different parts of a design to use different simulation modes. For example, a critical part of a design might be simulated at the transistor level while another part is simulated at the functional level.

10.3.1 Stages for simulation in design flow

Simulation is used at multiple stages during the design process. Various simulation stages are shown in figure 10-2. Initial prelayout simulations include logic-cell delays but no interconnect delays. Estimates of capacitance may be included after completing logic synthesis, but only after physical design is it possible to perform an accurate postlayout simulation. Functional Simulation can be performed at following points in design:

- RTL simulation allows you to verify and simulate a description at system or chip level where system or chip is described using high level RTL language. Testbench is created to model the environment of the system or chip.

- Once RTL is bug free design is synthesized using vendor library and statistical wire load models. Timing constraint information is written out for floorplanner from synthesis. Post-synthesis gate-level simulation can be performed on VHDL or Verilog netlist. Testbench can be used again to stimulate and simulate the synthesized results and check its consistency with original design description.

- After synthesis, Floorplanner plans the overall placement that can meet timing constraints for synthesis. Partial timing functional simulation can be performed on this Post-map or floorplan netlist by back annotating SDF file from implementation after floorplan/mapping, and before place and route. The design does not contain net delays since it is not routed.

- After floorplan, place and route is performed, delay values and parasitic estimated values are again fed to floorplanner. Wire load models are back annotated into synthesis tools and design has to be analyzed at this point if it meets timings. Several iterations are generally need to finally close on timing.

- From final place and route, final delay and parasitic values are extracted to be used for static timing analysis or RTL-gate simulations as final verification. At this point simulation is done on final netlist simulation with full timing after worst case placed and routed delays are calculated for your design. You should be able to use the same testbench that was created in RTL simulation to verify gate level netlist.

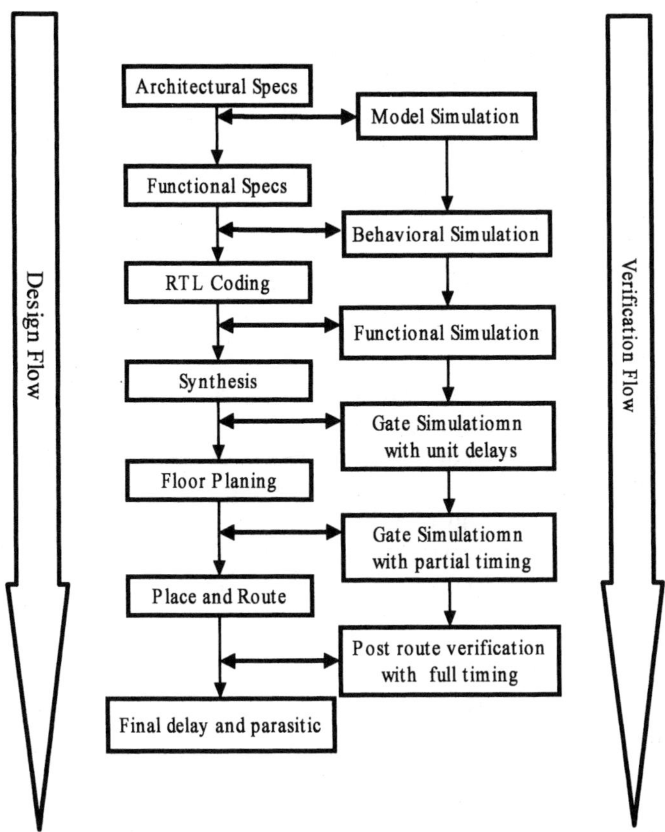

Figure 10-2. Simulation stages

10.4   Getting ready for gate simulation

10.4.1 Stimulus generation

The generation protocol should have timing information. The driver TVMs mentioned in Methodology chapter 6 should have the timing according to the protocol.
It is important that during gate simulation "XX" is driven on the bus beyond the
timing window to be able to check any setup or hold violations.

10.4.2 Back annotating SDF files

For designs that are multimillion gates, SDF files tend to be huge in giga bytes size. First challenge is to be able to successfully back annotate these in case chip level simulation is to be run with SDF back annotation. No tools can support back annotating of this huge SDF files. One of the workarounds is to reduce the size of the SDF files to remove the timing information that is redundant for running gate level simulations and that is more required for silicon. Some of the tools have option to dump the SDF files out that can merge the timing information to reduce the size of SDF files for gate simulations.

So before using the SDF files in gate simulation, some processing has to be done on these files because of their huge size. For example:

- \The sdf files generated by some tools, have parallel clock, reset buffers for every flop. For a multimillion chip there can be 100,000 flops. These can be replaced with one buffer. In the silicon these buffers are required but in simulation if these buffers are used with timing information there can be several glitches for example if different buffers are off by even few pico seconds. A script can be written to replace them with one buffer.
- Remove all "\" generated by tools since most of the simulators do not like it e.g. \[\sdf[].
- Remove any conditions.
- Flatten hierarchy - replace "/" with "--" the way some of the synthesis tools dumps the netlist.
- Some of the SDF files have -ve hold time, but simulators cannot take it. So, it requires some processing of SDF files to add -ve hold to buffer delays.
- If recording the waveform, make sure problems are set correctly and only to the signal levels that are to be recorded. probe should have option of stopping when it sees cells and not all the way down in the hierarchy.

10.4.3 Multicycle path

Mostly designs have many multicycle paths defined. These are the paths that need more than one clock cycle for becoming stable. These are found in designs that

- Use both positive and negative edge of clock.
- Use tristate buses.

- Also, when output of one flop is not immediately sampled in the next clock by another flop, instead output is sampled after two or more clocks, then it is declared as multicycle path. This is done to avoid unnecessary constraint on synthesis as well as place & route tools

For Static timing analysis, declaring a multicycle path can tell analyzer to adjust the measurements so it does not incorrectly report setup violations. They require some special multicycle setup and hold time calculations.

Some designs have thousands of multicycle paths. There is a chance of a multicycle path that is wrongly defined for single cycle path. This kind of problem can only be caught in the gate level simulations.

10.4.4 Using Sync Flops

Some designs use multiple asynchronous clocks. There are several critical signals that must cross clock domains in the design. These need valid synchronization based on flip flops, domain enabled multiplexor logic, FSMs and FIFOs. For example:

always @ (posedge clock1)

data_a <= Data_in;

always @(posedge clock2)

data_b <= data_a;

Where clock1 and clock2 are asynchronous. Signal data_a will be sampled to data_b at rising edge of clock2. Now when signal data_a will change at rising edge of clock2, there will be clock2 setup/hold violation and signal data_b will be unknown.

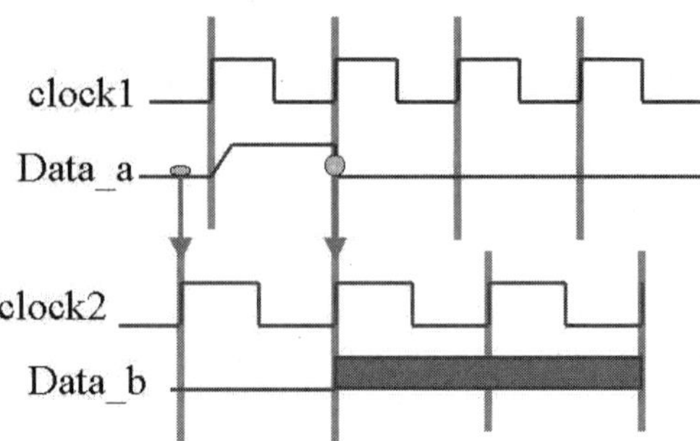

It is required to instantiate synchronization flops in such design which has signal that cross clock domains. It is important that during gate level simulation with SDF back annotation, all the synchronization flops are specified for no timing check. If you do not identify and set all the sync flops, you will get "x" in simulation results and have to run gate level simulation all over again. When fast, typical and slow simulations are run, delays through different clock trees vary differently. As a result a signal may start at sending clock domain as valid data, but it becomes "X" as it received by synchronization flop incase setup and hold timings are not met for the signal. Critical synchronization flops can be substituted as a special gate level simulation flop which never outputs X unless there is an X at the input. For valid 1'b0 or 1'b1 as input, there will always be valid 1'b0 or 1'b1 at its output.

In simulators, for gate level simulation there is an option to turn off the timing in specific parts of the design by using timing file. The option for ncsim is +nct-file+option for example:

ncverilog +nctfile+myfile.tfile source_files

While annotating with SDF file, design gets annotated using SDF file information, but the timing constructs that are specified in the timing file are removed from the specified instances.

10.4.5   ncpulse and transport delays

Timing simulation models the internal delays that are present in the real circuits. It is necessary to set the options for delays correctly while running gate level simulations. Else, each run takes considerable amount of time to run and to debug the problem in propagation of data to the outputs. This is illustrated by example below.Usually two models of delays are used

·   Inertial delays

- Transport delay

In inertial delay model when component has some internal delay and input changes for time less than that delay, no change is propagated to the output. Most often inertial delay is desired but sometimes all changes in input should have effect on the output. In transport delay model, short pulses are to absorbed as in inertial delay model. Inertial delays are easy to implement in simulator.

In the NC-Verilog simulator used in the design example, both module path delays and interconnect delays are simulated as transport delays by default. It is required to set pulse control limits to see transport delay behavior. If pulse control limits are not set, the limits are set equal to the delay by default, and no pulses having a shorter duration than the delay will pass through. That is, if pulse control limits are not set, module path delays and interconnect delays are simulated as transport delays, but the results look as if the delays are being simulated as inertial delays.

Different simulators have different options or default delay models. In Verilog-XL, delays are inertial by default. Following command-line plus options are used together

+transport_int_delays and +multisource_int_delays, to enable transport delay behavior with pulse control and the ability to specify unique delays for each source-load path.

The other option that can be used is -intermod_path option with elaborator (ncelab) to enable the ability to specify unique delays for source-load paths. By using this option, delays can be imposed from the same source to different loads, and different delays and pulse limits can be specified on paths from different sources to the same load.

10.4.6 Pulse handling by simulator

By default, pulse control limits are set to the delay to yield inertial delay behavior. To see transport delay behavior, pulse control limits are to be set:

- Use the -pulse_r and the -pulse_e options when you invoke the elaborator. These options set global pulse limits for both module path delays and interconnect delays. In order to set pulse control for module path delays and interconnect delays separately in the same simulation, following options can be used:
    -pulse_r and -pulse_e to set limits for path delays

-pulse_int_r and -pulse_int_e to set limits for interconnect delays.

- Simulator takes one of the following actions upon setting the global pulse control options:

    · Reject the output pulse (the state of the output is unaffected).

    · Let the output pulse through (the state of the output reflects the pulse).

    · Filter the output pulse to the error state. This generates a warning message and then maps to the x state.

The action that the simulator takes depends on the delay value and a window of acceptance. The simulator calculates the window of acceptance from the following two values that you supply as arguments to the options. Both arguments are percents of the delay.

- reject_percent

- error_percent

Syntax:

-pulse_r reject_percent -pulse_e error_percent (for elaborator)

or

ncpulse_r/args ncpulse_e/args for simulator

Using the reject limit and error limit calculations, the simulator acts on pulses according to the following rules:

- Reject if 0 <= pulse < (reject limit).

- Set to error if reject limit <= pulse < (error limit).

- Pass if pulse >= error limit.

10.4.6.1 Example

One of the initial gate simulation setup problem problems encountered in the design were related to setting up these correctly. For first simulation run the data was not seen all the way out of the SPI4 interface. This was because For SPI4 interface, delay of about 1.3ns through the LVDS Pads was seen(for tdat, tdclk, tctl). The values in the SDF is shown below,

(CELL

 (CELLTYPE "LDLVDS2_15T_TX")

(INSTANCE tdat_inst0)

(DELAY (ABSOLUTE

(IOPATH DIN TXN (1305:1305:1305) (1308:1308:1308))

(IOPATH DIN TXP (1308:1308:1308) (1305:1305:1305))

))

)

When the SPI4 Interface works at 400Mhz, the pulse width to these Pads are less than 1.25ns which is less than the delay through the pad. In the verilog simulation it was seen that only the input to the Pads toggle output did not toggle. Pulse width through the LVDS pad (1.25ns) can be less than the propagation delay (1.3 ns). in Silicon, the LVDS output will also be a pulse, rather than a constant as in simulation. The fix for this problem is to set appropriate parameter in Simulation so that LVDS can propagate correctly. So, following options were set in gate level simulations which could pass any small glitch.

+ncpulse_r/0 +ncpulse_e/1

10.5   Setting up the gate level simulation

When gate level netlist is ready after synthesis or ECO fix, or Place and route or any other process in the design flow and there is plan for running simulation on this, it is always helpful to first run simulation without any timing. It can help resolve and fix at least half of the setup issues related to gate simulation setup like following:

- The testbench should be able to generate stimulus, monitor results and check without any alterations. Still, if there are any holes or adjustments needed because hierarchy of some signals have changed for monitors, checking or assertions, it can be fixed.
- Gives extra confidence on the functionality of the netlist.

- Any machine related issues and tool related issues with gate simulation can be taken care of.

- Memory or chip configuration setup if different for gate simulation can be taken care of.

Simultation testbench should have option of being able to run gate simulation at various clock frequencies. In other words, testbench should be independent of clock frequency. This is required to generate functional vectors at different speeds and to debug any issues regarding the full speed simulation.

- Once the environment is setup, it is required to select the tests that should be run in gate simulation regression. Even when environment is setup and chip configuration is ready, each test might take 15 times more to run. For example test that is able to send in one hundred cells in five minutes on RTL simulation will probably take one hour once all the netlist is compiled and no timing information is backannotated

- Once functional simulation is running with gate netlist, it is time to annotate the SDF files. The problem with SDF files is normally they are huge. If you have the understanding of the design, you can reduce the size of SDF files significantly and get rid of redundant timing information in there for repeated clock buffers.

- Before annotating the SDF file, it should be compiled. Once SDF file is compiled, always use the compiled version of SDF files in simulation instead of recompiling it every time. Different tools have different setup command for this.

- All the warnings related to SDF files can be checked at the time of compilation, so while running simulation suppress all these warnings. For example ncsim has option "+sdf_nowarnings" .

- Specify the delays values that should be annotated - MINIMUM, TYPICAL, MAXIMUM. Generally sdf command file is written with all these options. Here is an example:

// File example.sdf_cmd

COMPPILED_SDF_FILE = "block1.sdf.X",

SCOPE: = block1.i1,

LOG_FILE = "sdf_block1.log",

MTM_CONTROL = "MINIMUM",

- List out all the Sync flops in the design to set a no timing check file

- List down what all warnings you want to suppress after the first run. For example you can use the command "+nowarn warning_code" to suppress warnings with specified warning code.

- Example of command line options added for running gate simulation is as below(ncsim simulator):

  +transport_int_delays

  +multisource_int_delays

  +maxdelay

  +nowarnTFNPC

  +nowarnTFMPC

  +nowarnIWFA

  +nctfile+./gate_files/no_timing_check.tfile

  +ncpulse_r/0 +ncpulse_e/1

  +sdf_nowarnings

- If all the above steps are done right, the expected results might come out during first run. Test run time is a lot during gate simulation with backannotation, so it is better to do thorough review of all the setup options chosen for gate simulation to avoid debugging by running the tests. If SDF file is back annotated, mostly the problems are seen with the missing information in terms of sync flops, information mismatch between SDF files and the timing library in terms of naming etc. because of tool issues. In all of these cases the SDF files have to be regenerated with different options or some post processing with the scripts is required.

It is a good idea to setup different databases for gate simulation at different speeds for generating functional vectors or debuging timing issues. Because once the simulation setup is done, test run does not take a lot of time. Mostly, gate simulations complexity is involved with first time setup of the environment, compilation of netlists and annotation on SDF files.

10.6   ATE vector Generation

10.6.1   Requirement for Functional testing on ATE

In addition to SCAN, Memory Bist and JTAG, it is preferable to run few functional vectors in ATE for following reasons

- To screen parts based functional vectors pass/fail.

- To make all inputs and outputs to toggle for characterization purpose.
- Also, running ATE at speed will verify that part meets target frequency.

10.6.2   Various ATE Stages

10.6.3   Choosing functional simulations for ATE

This step involves going through functional test plan and then choosing ones which covers major part of logic. Once these functional simulations are chosen, run functional simulation on RTL and make necessary changes so that most part of silicon is exercised. Run the simulation and verify the results. Like this, select few more relevant functional vectors.

10.6.3.1   Running gate level simulations

Run gate level simulations with extracted delay information from SDF file. When simulations are run with SDF, it reflects actual gate delays. SDF in general is extracted for 3 different cases of process.

- Best case: When process has minimum delay, temperature minimum and voltage maximum.
- Typical case: Normal process delay. Room temperature and normal voltage
- Worst case: Worst delay due to process, minimum voltage and maximum temperature allowed.

For all three simulations results should be same.

If the result does not match it could be due to various facts which were not taken care during design.

10.6.3.2   Typical mistakes designers make are

- Clock is not stable when reset is removed. This happens if logic (PLL or DLL) also reset using reset pin
- Some flops can not be reset. These flops can come up with any value and cause output to mismatch. Ideally all flops should be reset.
- Logic across different clock domain need to meet timing assuming all clocks are multiple of master clock. This way it can be guaranteed that if all external clocks are derivatives of master clock, logic across different clock domain meet

timings and result will always match. Otherwise logic may be working but outputs may be seen at different clocks because of synchronizing logic.

· Logic should not generate any clock internally.

If one or more of above is not taken care in design, it might result in unpredictable output timings. It may be required to adjust input clocks and signals with respect to each other to achieve predictable output timings. In some case it may require to reduce target frequency.

10.6.4 Generating vectors

From waveform data base it is required to convert them to vectors based on the format required by tester. Different testers need different input formats. Most of tester takes. WGL files and some need vendor specific formats. Using script they can be converted to 1s and 0s for each input at each master clock edge. Most of the tester works on one master clock and they need inputs to toggle with respect to master clocks. Ideally these different vectors need to be generated for each test,

· Best case.
· Typical.
· Worst case.

10.6.5 Applying vectors back to netlist

In standard test bench these vectors should be applied back to netlist and verify output from vectors and simulation match. This step is required to make sure when these vectors are used in tester they give same result as simulation. Since tester time is very expensive this step is required to cleanup vectors before run on tester.

There could be various reasons why applying vectors back to netlist may not pass first time. In actual simulation you may not be able to control input as per timing requirement. This can be easily done when applying vectors back to netlist. Control each input as per data sheet requirement. This will guarantee that less time will be spent at tester.

10.6.6 Testing at ATE

After successful testing at simulation environment vectors are ready to be tried at ATE.

This requires

- Board with socket for DUT, built for this purpose with all power and grounds connected properly.
- Vectors converted from simulation to the format tester supports.

It is always recommended to try vectors at lower frequency to sort out logical issues. It requires thorough understanding of the product to debug at tester. When results are not seen as expected it is required to analyze output pins at each stage. Depending on where first output failed, need to analyze and understand why this first output is not seen. In most of the cases this may be due to not setting timing information properly at the inputs. Once these things are sorted out, most of functional vectors pass without any issues.

10.6.7 Problem at speed vectors

If DUT is working with just one clock there may not be many issues. But most of today's ASICs work on multiple clocks and double data rate on most of the clocks. This further complicates testing. Most of testers available work frequencies less than 300MHz and generally work on rising edge clock. So some compromises to be made depending tester used.

10.6.7.1 Characterizing the product

Other than running vectors and using for production screening, these vectors are used to characterize the product. Step involved to characterize

- INPUT window: All inputs needs to be tested for valid window as required. Each input setup time and hold time is characterized by toggling outside setup+hold window. Any toggle outside this window should not change the output. Also input voltage levels are tested as per data sheet requirements.
- Outputs delays: All outputs are measured across voltage, temp and process range to check if output delay timing requirements are met. Output delays are also measured by varying loading at the output. This can be easily achieved as testers support programmable loading for outputs.

10.7   Setting up simulation for ATE vector generation

Simulation needs to be setup to be able to generate vectors at any clock frequency. Vectors can be generated by writing a wrapper around the chip testbench that dumps vectors while simulation is running in the format required. -

- Pinlist for all input output interfaces will be required and wrapper should be able to dump the vectors to the files in the required format.

- Some renaming or reassigning will be required if netlist has different names for buses. For tester PADRING name is printed while the wire name in RTL code can be the old name.

- Different conventions are followed to print to the files and convert the signals to the tester type signals.

```
/* input enable test_type
    0    1    0
    z    1    Z
    1    1    1
    0    0    L
    z    0    Z
    1    0    H
*/

   function [7:0] test_type;
   input signal; // signal to be converted
   input enable; // 1 = output mode, 0 = input mode

   if ( enable == 1'b0)
     begin
       if (signal === 1'b0)
         test_type = "0";
       else if (signal === 1'b1)
         test_type = "1";
       else if (signal === 1'bz)
         test_type = "z";
       else
         test_type = "x";
     end
   else
     begin
       if (signal === 1'b0)
         test_type = "L";
       else if (signal === 1'b1)
```

```
        test_type = "H";
      else if (signal === 1'bz)
        test_type = "z";
      else
        test_type = "x";
   end

 endfunction
```

- Simulation should be setup to run at different clock modes

10.8   Examples of some issues uncovered by gate simulations

- Spec mentioned number of system clocks required for read, write and recovery timings. These timings worked perfectly for RTL simulation. But gate level simulation was done with extracted worst case timing info, it was found that these numbers were not sufficient. This called for change in spec and in turn change in some customers programming.

- Output of two gates ANDed and used as clock to some delay elements. It was expected that only when both are 1, output should be 1. In one instance gate A output was going H to L and gate B was going L to H. In layout it was expected that gate A delay is shorter so that output goes low before gate B switches. Since these constraints were not met, it happened that for a short time there was glitch at the output of AND gate. This was discovered only in gate level simulation.

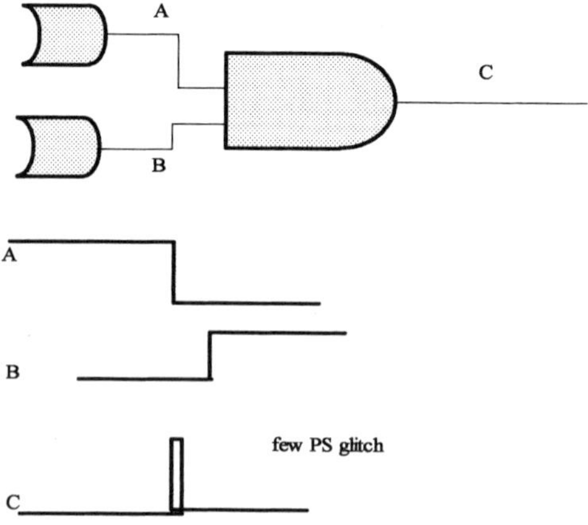

- There were wrongly declared multicycle paths in the design. During timing analysis this path was ignored as it was mentioned as 2 clocks from Flop A o/p to Flop B i/p. Since there are no timing info in RTL it was never discovered in RTL simulation that this path is indeed not MCP but single cycle path. During gate level simulation with timing info, it was discovered that this is Single cycle path.

APPENDIX

The appendix gives example code for simple interfaces implemented based on the methodology discussed in previous chapters. It also has code for the standard interfaces PL3 RX, PL3 TX and SPI4.

0.1   Common Infrastructure

In the commmon infrastructure were defined objects that represent networking data strucutures, network protocols, common generation, checker and monitor classes. Here are some the main classes defined in the common Infrastructure.

- tbvDataCom: Defines the DataCom class derived from tbvSmartRecordT. For different dataCom types it has data access methods to return Data, Size, header, Payload, trailer, get data by 16, 32 or 64. Set payload, header size, insert bytes, check to see if it is a packet or a cell. Also defines control data access methods, to get or set datacom type, control word, pid, Qos, SN, SOP, EOP, SOB, EOB, CLP, EFCI, Valid, Bad, OAM.

- tbvCell: Defines tbvCellT class derived from tbvDataComT. It implements cell data structure. Defines various functions to extract from cell: Port , Qos, SN, SOP, EOP, EOB, SOB, CLP, EFCI, Valid, Bad, OAM.

- tbvINTCell: Defines tbvINTCellT for internal cells from various blocks, derived from tbvCellT. Various access methods for accessing cell Type, Payload, Pad Size and for setting or creating payload.

- tbvATMCell: Defines ATM cell class tbvATMCellT derived from tbvCellT. Implements ATM cell data structure. Various methods for access from header fields to set and get GFC, VPI, VCI, PTI, NNI. Also to set and get Header, Payload.

- tbvSwitchCell: Defines tbvSwitchCellT class derived from tbvCellT. Access functions to get header, type and header size, and to set header , headersize.

- crcFunctions : Functions to calculate crc32, crc16 and crc10.

- tbvPkt: Defines tbvPkt class derived from tbvCellT. It implements packet data structure.   Various functions to access control fields.

- tbvEtherPkt:Implements Ethernet packet data structure.

- tbvMPLS: implements network MPLS tag data structure.

- tbvATOMControlWord: Defines ATOM(ATM) Control word. Implements functions for setting and accessing transport type, EFCI, CLP, Command Response, Length, Sequence number, Control word.

- tbvSar:Defines tbvSarT class which has segmentation and reassembly functions to convert from packet to cell/ATM cell and from cell to packet.

- tbvSwitchPkt: Implements Switch packet data structure.

- tbvTrafficGen : Implements traffic generator. Can set leaky bucket param for single leaky bucket or dual leaky bucket. Actual rate can be converted to leaky bucket parameters. Flowmem holds all the information about flow, qos and port. Use a tbvSmartQueue to store the cells to be sent to. Hash map used to store the leaky bucket for each flow id.

- tbvCheckerTvm: It implements a generic checker base class which can be used by all the different chip or block checkers. The individual checker only need to derive from the base checker. Then block specific process function will override the base class's process function to perform the specific block requirments. The base class provide the basic functions such as receiving data, checking data, sending data, link the predecessor with the successor, It also records the statistics. It provides help function for crc check, sop check, eop check. aal5 trailer extraction. The input data is processed and stored by an index selected by tbvCheckerQueueSelector to a smartQueue. The index can be port, flow, qos or anything.

- tbvMonitorTaskT:Base class for all the monitors.

0.2   Simple example based on above methodology

This section discusses testBuilder code for reassembly block of the chip. Block specific code is provided here for clarification of concepts. Common code from the common infrastructure is not provided. The code shared here is not showing any block specific algorithms, it is showing a dummy behavior where dataGen generates one traffic type through driver TVM.

NOTE: The whole executable example will be provided either with CD or on the website. It will be too much code to discuss here otherwise.

0.2.1   Generator

Common code for the generator is implemented in the common infrastructure. RAS block data generator inherits from the common data generator. Here is the common code that is used for block level:

dataprovider: /** \ingroup trafficGenerator data generator. This class upon request generates patterns and puts into a queue. The pure virtual function is provided, that must be implemented at the block level that provides implementation of instructing the generator to generate certain pattern of data on fids. trafficSchedulerTask's job is to request for a cell for a given fid when is scheduled to send and this object provides cells until a pre set number of cells are generated.*/

dataGram Class: /** \defgroup monitorDriver monitor and driver related class This is the base class for self sufficient entity class (meaning data as well as control information). The sub class will implement data and control related functions. For example, the rasDataGramT should inherit from this, implement data and control. Basically it should have a pointer to dataPayloadT(for carrying payload) and a pointer to controlDataT(for side band info). The interface monitor/driver should use interface specific instance of dataGram(rasDataGramT).*/

Following code is specific to the block:

rasblockDataGen.h
```
#ifndef RAS_DATA_GEN
#define RAS_DATA_GEN

#include "common/dataProvider.h"
#include "common/dataGram.h"
```

```cpp
#include "../sideband/ras/MemRasSidebandControl.h"
#include <stl.h>

/* \ingroup generator reassembly specific data generator
   This class inherits dataProviderT, implements pure virtual function
   list<dataGramT *> * getListOfCells(dataGramTT dataGramP).
   it supplies to tbvTrafficGenTaskT.

*/

class reassemblyDataGenT : public dataProviderT {
 public:
reassemblyDataGenT();
  ~reassemblyDataGenT();

  bool setupTraffic(memRasSidebandControlT * sidebandControl);
  void addDataGramRef(dataGramT * dataGramP,unsigned int qid,
      unsigned int num = 1);
  tbvBagT<dataGramT *> getDataGramRef(unsigned int fid);
  void removeDataGramRef(dataGramT * dataGramP,unsigned int qid);
  bool rotate(const dataGramT * dataGramP);
 protected:
 private:
  list<dataGramT *> * getListOfCells(unsigned int fid);
  void connectGenerator(dataGramT * dataGram,float perQidWeight = 1.0,
float perPortWeight = 1.0);
  void setup();

};

#endif
```

rasblockDataGen.cc

```cpp
#include "rasBlockDataGen.h"

reassemblyDataGenT::reassemblyDataGenT()
{
  setup();
}

/*****************************************************************************/

reassemblyDataGenT::~reassemblyDataGenT(){
}

/*****************************************************************************/

list<dataGramT *> * reassemblyDataGenT::getListOfCells(unsigned int fid){

  dataGramRefBagMutex.lock();
  tbvBagT<dataGramT *> &dataGramRefBag = dataGramRefs[fid];
  dataGramT * dataGramRefP;
  dataGramRefP = dataGramRefBag.peekNext();
```

```
    dataGramRefBagMutex.unlock();
    dataGramRefP->getDataComP()->randomPayload();
    totalNoOfPacketsSent++;

    list<dataGramT *> * cells = new list<dataGramT *>();

    dataGramT * outputDataGram = new dataGramT();
    dataComT * dataCom = dataGramRefP->getDataComP();

    dataPayloadT * payload = new dataPayloadT(dataCom->getDataAsBytes());
    //duplicate control data
    sidebandControlT * sidebandControl = (dataGramRefP->getSidebandControlP()->duplicate());
    memRasSidebandControlT * rasSidebandControl = ((memRasSidebandControlT *) sidebandControl);

    outputDataGram->setDataComP(payload);
    outputDataGram->setSidebandControlP(rasSidebandControl);
    cells->push_back(outputDataGram);
    return cells;
}


/**************************************************************************/
/// return false if setup failed.
bool reassemblyDataGenT::setupTraffic(memRasSidebandControlT * sidebandControl){
    bool success = true;

    switch(sidebandControl->mem_ras_type){
    case 0: {
      dataPayloadT * dataPayload = new dataPayloadT(64);
      dataGramT * dataGram = new dataGramT(dataPayload,sidebandControl,"memRasTraffic");
      connectGenerator(dataGram);
      break;
    }

    default:{
      success = false;
      break;
    }
    }

    return success;
}


/**************************************************************************/

void reassemblyDataGenT::connectGenerator(dataGramT * dataGram,
      float perQidWeight,
      float perPortWeight) {
    memRasSidebandControlT sidebandControl =
      *((memRasSidebandControlT *) dataGram->getSidebandControlP());

    getFlowMem().linkFidToPort(sidebandControl.mem_ras_qid,
      sidebandControl.mem_ras_portid,perQidWeight);
    getFlowMem().setPortParam(sidebandControl.mem_ras_portid,perPortWeight);
    addDataGramRef(dataGram,sidebandControl.mem_ras_qid);
}
```

```cpp
/*******************************************************************************/
void reassemblyDataGenT::addDataGramRef(dataGramT * dataGramP,unsigned int qid,
  unsigned int num){
 dataGramRefBagMutex.lock();
 memRasSidebandControlT sidebandControl =
  *((memRasSidebandControlT *) dataGramP->getSidebandControlP());
 dataGramRefs[qid].add(dataGramP, num);
 dataGramRefBagMutex.unlock();
}
/*******************************************************************************/
tbvBagT<dataGramT *> reassemblyDataGenT::getDataGramRef(unsigned int qid){
 return dataGramRefs[qid];
}
/*******************************************************************************/
void reassemblyDataGenT::removeDataGramRef(dataGramT * dataGramP,unsigned int qid){
 dataGramRefBagMutex.lock();
 memRasSidebandControlT sidebandControl =
  *((memRasSidebandControlT *) dataGramP->getSidebandControlP());
 unsigned int port = sidebandControl.mem_ras_portid;
 dataGramRefs[qid].resetPeek();
 dataGramT * peekP = dataGramRefs[qid].peekNext();
 unsigned int bagsize = dataGramRefs[qid].size();
 while(peekP != dataGramP && bagsize != 0) {
  peekP = dataGramRefs[qid].peekNext();
  bagsize--;
 }
 if(bagsize != 0) {
  dataGramRefs[qid].remove(TRUE);
 }
 dataGramRefs[qid].resetPeek();

 if(dataGramRefs[qid].empty()) {
  flowMemT & flowMem = getFlowMem();
  flowMem.delinkFidFromPort(qid,port);
 }
 dataGramRefBagMutex.unlock();

}
/*******************************************************************************/
bool reassemblyDataGenT::rotate(const dataGramT * dataGramP){
 // rotate traffic on port if dataGramP has eop.
 cout<<"from reassemblyDataGenT::rotate"<<endl;
 memRasSidebandControlT sidebandControl =
  *((memRasSidebandControlT *) dataGramP->getSidebandControlP());
 bool rotate = false;
 if(sidebandControl.mem_ras_eop == 1)
  rotate = true;

 return rotate;
}
/*******************************************************************************/
void reassemblyDataGenT::setup(){
}
```

0.2.2 Sideband

The common code used here is again from dataGram.h and controlData.h.

controlData.h: /** \defgroup control \ingroup control data structure, Base class for the control data structure. All side band and inband. The inherited control(for a given interface) must be bundled together with a dataPayloadT instance. This makes a pair for carrying sufficient information for a given interface. For example, Input to Ras say RasInputCell should have pointer to dataPayloadT and pointer to rasInputControl(must inherit from this class). */

memRasSidebandControl.h

```
#ifndef MEM_RAS_SIDEBAND_CONTROL_H
#define MEM_RAS_SIDEBAND_CONTROL_H

#include "common/controlData.h"
#include "common/dataGram.h"

/** \ingroup control side band control signals for reassembly.
    This class inherits from sidebandControlT and implemets
    reassembly(mem->ras) specific side band signal.

    RAS input monitors/drivers and checker updates(sets its variables) instance of this class.
    Instance of this class will be bundled in instance dataGramT and test bench's interblock communication
    will be through instance of dataGramT.*/


class memRasSidebandControlT : public sidebandControlT {
public:
  memRasSidebandControlT(const char * nameP = NULL);
  ~memRasSidebandControlT();

  virtual memRasSidebandControlT * duplicate();
  memRasSidebandControlT & operator=(const memRasSidebandControlT & rhs);
  unsigned int   reset_l;
  unsigned int   clk;


  unsigned int   mem_ras_available;
  unsigned int   mem_ras_data;
  unsigned int   mem_ras_eop;
  unsigned int   mem_ras_portid;
  unsigned int   mem_ras_sop;
  unsigned int   mem_ras_type;


  unsigned int   ras_mem_ctrl_pop;
  unsigned int   ras_mem_data_pop;
protected:
```

```
private:
 memRasSidebandControlT(const sidebandControlT & rhs,const char * nameP = NULL);

};
#endif
```

memRasSidebandControl.cc

```
#include "memRasSidebandControl.h"

memRasSidebandControlT::memRasSidebandControlT(const char * nameP) :
 sidebandControlT(nameP){

}

memRasSidebandControlT::memRasSidebandControlT(const sidebandControlT & rhs,
    const char * nameP){
 operator=(rhs);
}

memRasSidebandControlT * memRasSidebandControlT::duplicate(){
 return new memRasSidebandControlT(*this);
}
/*****************************************************************************/

memRasSidebandControlT & memRasSidebandControlT::operator=(const memRasSidebandControlT & rhs){

 sidebandControlT::operator=(rhs);
 this->reset_l = rhs.reset_l;
 this->clk = rhs.clk;
 this->mem_ras_available = rhs.mem_ras_available;
 this->mem_ras_data = rhs.mem_ras_data;
 this->mem_ras_eop = rhs.mem_ras_eop;
 this->mem_ras_portid = rhs.mem_ras_portid;
 this->mem_ras_sop = rhs.mem_ras_sop;
 this->mem_ras_type = rhs.mem_ras_type;
 this->ras_mem_ctrl_pop = rhs.ras_mem_ctrl_pop;
 this->ras_mem_data_pop = rhs.ras_mem_data_pop;
 return *this;
}
/*****************************************************************************/
memRasSidebandControlT::~memRasSidebandControlT(){

 //clean up any memory allocated

}
/*****************************************************************************/
```

0.2.3 Driver

Driver will have ".v" file to show connectivity of C++ Testbuilder TVM with the RTL code. Common code this shares is the interfaceDriver task, dataGram, dataPayload and messageResponder.

interfaceDriverTask:

/** \ingroup monitorDriver base class for all interface drivers. Base class for all interface driver classes. Inherited classes are expected to implement body(). This class takes care of recording transactions into fiber, starting and stopping of thread. In addition there is an interfaceDriverQueue (tbvMailbox), the data gram to be driven must be pushed into this queue */

messgeResponder and message task are already discussed in detail.

memRasDriver.v

```verilog
module memRasDriverTvm(
          reset_l,
          clk,
          cpu_ras_cs_l,
          cpu_rdwr_l,
          cpu_addr,
          cpu_data_in,
          mem_ras_available,
          mem_ras_data,
          mem_ras_eop,
          mem_ras_portid,
          mem_ras_sop,
          mem_ras_type,
          ras_mem_ctrl_pop,
          ras_mem_data_pop,
     );

   input         reset_l;
   input         clk;
   output        mem_ras_available;
   output [127:0]  mem_ras_data;
   output        mem_ras_eop;

   output [6:0]    mem_ras_portid; // incoming cell's port id
   output        mem_ras_sop;  // start of packet
   output [3:0]    mem_ras_type;  // cell type
   input         ras_mem_ctrl_pop;  // ctrl pop is used for rest of epe ctrl
   input         ras_mem_data_pop;  // data pop is used with data & byte_en
   input [5:0]    cpu_addr;
   input [31:0]   cpu_data_in;
   input         cpu_ras_cs_l;
   input         cpu_rdwr_l;
   reg ras_pop;
   // These registers are driven from C++
```

```
 reg         mem_ras_available_o;
 reg [127:0]  mem_ras_data_o;
 reg         mem_ras_eop_o;
 reg [6:0]    mem_ras_portid_o;
 reg         mem_ras_sop_o;
 reg [3:0]    mem_ras_type_o;
// These assignments create Verilog drivers for the C++ regs
 assign mem_ras_available  =    mem_ras_available_o;
 assign mem_ras_data       =    mem_ras_data_o;
 assign mem_ras_eop        =    mem_ras_eop_o;
 assign mem_ras_sop        =    mem_ras_sop_o;
 assign mem_ras_type       =    mem_ras_type_o;
 //
 //  register every instance of this TVM to test builder
 //

initial
 begin
     mem_ras_available_o  = 0;
     mem_ras_data_o       = 0;
     mem_ras_eop_o        = 0;
     mem_ras_portid_o     = 0;
     mem_ras_sop_o        = 0;
     mem_ras_type_o       = 0;

 end

 initial
   $tbv_tvm_connect;
endmodule
```

memRasDriver.h

```
#ifndef MEM_RAS_DRIVER_H
#define MEM_RAS_DRIVER_H


#include "TestBuilder.h"
#include "common/interfaceDriverTask.h"
#include "common/dataGram.h"
#include "common/dataPayload.h"
#include "../sideband/memRasSidebandControl.h"
#include "common/message.h"
#include "common/messageResponder.h"
```

/** \ingroup monitorDriver memory to reassembly driver.
   It has one tvm(to connect to HDL signals of the design) called memRasDriverTvmT,
   and one task to drive signals called memRasDriverTaskT.

   The memRasDriverTaskT inherits from intefaceDriverTaskT.The siganls to be stimulated
   is passed through a mail box(tbvMailboxT ,which is a queue with semophore), which is
   defined in memRasDriverTaskT.

   The task is spawned only once and keeps waiting on mail box untill asked to stop.

```cpp
    Pre-requisits: The memRasControlT must be defined which will have all mem to ras
    side band control signals.
*/


class memRasDriverTvmT;


class memRasDriverTaskT : public interfaceDriverTaskT {
 public:
  memRasDriverTaskT(memRasDriverTvmT & tvm, const char * nameP = NULL);
  ~memRasDriverTaskT();
  memRasDriverTvmT & memRasInterface;
  void body ();
};


/***********************************************************************************************/

class memRasDriverTvmT : public tbvTvmT
{
 public:
  memRasDriverTvmT(const char * nameP = NULL);
  ~memRasDriverTvmT();

  //hdl interface signals
  tbvSignalHdlT   reset_l;
  tbvSignalHdlT   clk;

  tbvSignalHdlT   mem_ras_available;
  tbvSignalHdlT   mem_ras_data;
  tbvSignalHdlT   mem_ras_eop;
  tbvSignalHdlT   mem_ras_portid;
  tbvSignalHdlT   mem_ras_sop;
  tbvSignalHdlT   mem_ras_type;

  tbvSignalHdlT   ras_mem_ctrl_pop;
  tbvSignalHdlT   ras_mem_data_pop;

  memRasDriverTaskT memRasDriverTask;


  // This function is called from $tbv_tvm_connect, to Create a TVM object.
  static void create ( );
 protected:
 private:

};

#endif
```

memRasDriver.cc

```cpp
#include "memRasDriver.h"
```

```cpp
TVM_VERILOG_REGISTER(memRasDriverTvmT::create , "memRasDriverTvm" );

memRasDriverTaskT::memRasDriverTaskT(memRasDriverTvmT & tvm, const char * nameP):
 interfaceDriverTaskT(&tvm,nameP),
 memRasInterface(tvm)
{
}


/*******************************************************************************/

memRasDriverTaskT::~memRasDriverTaskT(){
}

/*******************************************************************************/
void memRasDriverTaskT::body(){

 //get dataGramT * from mail box,get memRasSidebandControlT and dataPayloadT from dataGramT.
 // Drive all mem to ras side band signals from memRasSidebandControlT and data from dataPayloadT.


 // debugOutput()<<"##############ras driver spawned @ time =  " << tbvGetInt64Time(TBV_NS) << "   "
 << running<<endl;

   tbvThreadT::yield();

   dataGramT * dataGram = popFromInterfaceDriverQueue();

   dataComT * dataCom = NULL;
   memRasSidebandControlT memRasSidebandControl;

   // The thread goes to sleeping state untill atleast one entry is there in mail box.

   // check whether dataPayloadT is NULL or not, if it is NULL,register message with message handler.
   if(dataGram->getDataComP() == NULL){
     tbvSafeHandleT<messageT> msg = (messageResponderT::createMessage());
     (*msg).setMessage(messageT::ILLEGAL_OPERATION);

     (*msg).setUserMessageString("NO DATA DETECTED:From mem to ras driver NULL poite in dataGram
 is detected");
     messageResponderT::report(msg);

     debugOutput()<<"ERROR: dataComp is NULL"<<endl;
   }else{
     dataCom = dataGram->getDataComP();
   }
   vector<uint64> data = getDataBy64(dataCom->getDataAsBytes());
   //Check whether memRasSidebandControlT is NULL or not, If it is NULL it is one of the fatal error
   // condition.
   if(dataGram->getSidebandControlP() != NULL){

   }else{
     memRasSidebandControl = *((memRasSidebandControlT *)(dataGram->getSidebandControlP()));
```

```cpp
}

  //Now we got side and as well data in memRasSidebandControl and dataPayload.

    while (tbv4StateNotEqual(memRasInterface.reset_l, 1)){
     tbvWaitCycle(memRasInterface.clk, tbvThreadT::POSEDGE);
    }

  debugOutput()<<"from Driver @ time = " << tbvGetInt64Time(TBV_NS) << "   " << *dataCom<<endl;
   memRasInterface.mem_ras_available = 1;
   for (unsigned int i = 0; i < data.size(); i++)
     {
// wait for one more cycle to put data on bus
memRasInterface.mem_ras_data(127,64) = data[i];
i++;
if (i <  data.size())
 {
   memRasInterface.mem_ras_data(63,0) = data[i];
 }
else
 {
   memRasInterface.mem_ras_data(63,0) = 0;
 }
tbvWaitCycle(memRasInterface.clk, tbvThreadT::POSEDGE);
tbvWaitCycle(memRasInterface.clk, tbvThreadT::POSEDGE);
     }
    delete dataGram;
    debugOutput() << dec << "@" << tbvGetInt64Time(TBV_NS) << ": Verilog data: RasDriver " << memRas-
Interface.mem_ras_data << endl;

   memRasInterface.mem_ras_available = 0;

}

/**************************************************************************************/
//                         Mem to Ras interface Tvm
/**************************************************************************************/
memRasDriverTvmT::memRasDriverTvmT( const char *nameP) :
 //tbvTvmT(),
 reset_l(getFullInterfaceHdlNameP("reset_l") ),
 clk(getFullInterfaceHdlNameP("clk") ),
 mem_ras_available(getFullInterfaceHdlNameP("mem_ras_available_o") ),
 mem_ras_data(getFullInterfaceHdlNameP("mem_ras_data_o") ),
 mem_ras_eop(getFullInterfaceHdlNameP("mem_ras_eop") ),
 mem_ras_portid(getFullInterfaceHdlNameP("mem_ras_portid") ),
 mem_ras_sop(getFullInterfaceHdlNameP("mem_ras_sop") ),
 mem_ras_type(getFullInterfaceHdlNameP("mem_ras_type_o") ),
 ras_mem_ctrl_pop(getFullInterfaceHdlNameP("ras_mem_ctrl_pop") ),
 ras_mem_data_pop(getFullInterfaceHdlNameP("ras_mem_data_pop") ),
 memRasDriverTask( * this, "ras Driver"  )
{

}

/**************************************************************************************/
```

```
memRasDriverTvmT::~memRasDriverTvmT(){
  //delete pointer if any
}
/************************************************************************************************/
void memRasDriverTvmT::create(){
  new memRasDriverTvmT();
}
```

0.2.4 Monitor

The monitor needs to deal with Veriolog RTL, so it also needs a verilog file to define the signal connectivity between Verilog RTL and C++ Testbuilder monitor TVM. The common code it shares is the common monitor Task.

monitorTask: /** \ingroup monitorDriver base class for all monitors. Abstract base class for all monitor classes. Inherited classes are expected to implement body() and allocate(); Argument of the body() is typesafed to dataGramT * .This class takes care of recording transactions into fiber, starting and stopping of thread. In addition there is a monitorQueue(tbvMailbox), the data read by monitor must be pushed into this queue;*/

memRasMonitor.v

```
module memRasMonitorTvm(
            reset_l,
            clk,
            mem_ras_available,
            mem_ras_data,
            mem_ras_eop,
            mem_ras_portid,
             mem_ras_sop,
          mem_ras_type,
            ras_mem_ctrl_pop,
            ras_mem_data_pop,
    );

    input          reset_l;
    input          clk;
    input          ras_mem_ctrl_pop;
    input          ras_mem_data_pop;
    input          mem_ras_available;
    input [127:0]  mem_ras_data;
    input          mem_ras_eop;
    input [6:0]    mem_ras_portid;
    input          mem_ras_sop;
    input [3:0]    mem_ras_type;
```

```
   //
   //  register every instance of this TVM to test builder
   //

   initial
    begin
    end
    initial
      $tbv_tvm_connect;

   endmodule
```

memRasMonitor.h

```cpp
#ifndef MEM_RAS_MONITOR
#define MEM_RAS_MONITOR

#include "TestBuilder.h"
#include "common/monitorTask.h"
#include "common/dataPayload.h"
#include "../sideband/memRasSidebandControl.h"
/** \ingroup monitorDriver memory to ras interface monitor.
   This file has a tvm(to connect HDL signal) and atleast one task to monitor
   memory to ras interface.

   The memRasMonitorT is inherited from monitorTaskT, implements allocate(),
   and body().

   Every time when a interface specific condition is met(when transaction is complete)
   this task pushes that info. via a pointer to dataGramT( *dataGramT).

   Everytime a new pointer to a object dataGramT is allocated via allocate() function,deletion of
   this pointer taken care by caller of this monitor.

*/

class memRasMonitorTvmT;
class memRasMonitorTaskT : public monitorTaskT {
 public:
  // The constructor must take tvm ref.
  memRasMonitorTaskT(memRasMonitorTvmT & tvm, tbvSignalHdlT &_clk, const char * nameP = NULL);
  virtual ~memRasmemRasMonitorTaskT(){}
 protected:
 private:
  void body();
  memRasMonitorTvmT & memRasInterface;

};

//memRasMonitorTvmT
/*****************************************************************************************************/
class memRasMonitorTvmT : public tbvTvmT {
 public:
   memRasMonitorTvmT(const char * nameP = NULL);
   virtual ~memRasMonitorTvmT();
```

```cpp
  // HDL signals at ras PE interface
  tbvSignalHdlT   reset_l;
  tbvSignalHdlT   clk;
  tbvSignalHdlT   mem_ras_available;
  tbvSignalHdlT   mem_ras_data;
  tbvSignalHdlT   mem_ras_eop;
  tbvSignalHdlT   mem_ras_portid;
  tbvSignalHdlT   mem_ras_sop;
  tbvSignalHdlT   mem_ras_type;
  tbvSignalHdlT   ras_mem_ctrl_pop;
  tbvSignalHdlT   ras_mem_data_pop;
  // Tasks in memRasMonitorTvmT
  memRasMonitorTaskT memRasMonitorTask;
 // This function is called from $tbv_tvm_connect, to Create a TVM object.
   static void create ( );
protected:
private:
};

#endif
```

memRasMonitor.cc

```cpp
#include "memRasMonitor.h"
TVM_VERILOG_REGISTER(memRasMonitorTvmT::create , "memRasMonitorTvm" );
memRasMonitorTaskT::memRasMonitorTaskT(memRasMonitorTvmT & tvm, tbvSignalHdlT &_clk,

 const char * nameP):
monitorTaskT(&tvm,_clk,nameP),
memRasInterface(tvm){

}


/*******************************************************************************************/
void memRasMonitorTaskT::body(){

  //Allocate new dataGramT pointer,(which has *dataPayloadT and *memRasSidebandControlT).
  //Populate memRasSidebandControlT members such as mem_ras_available, mem_ras_clp etc.
  //create dataPayloadT out of data on data bus(mem_ras_data),set *dataPayloadT and
  // *memRasSidebandControlT in dataGramT and push into mail box.
  debugOutput() << dec << "ras monitor spawned @ time =  " << tbvGetInt64Time(TBV_NS) << endl;

  tbvSignalHdlT &clk = getClk();
   dataGramT * dataGram = allocate();
   memRasSidebandControlT * memRasSidebandSignal =
    new memRasSidebandControlT("Mem to Ras side band control");
   dataPayloadT * dataPayload;
   while (tbv4StateNotEqual(memRasInterface.reset_l, 1)){
     tbvWaitCycle(clk, tbvThreadT::POSEDGE);
   }

   while (tbv4StateNotEqual(memRasInterface.mem_ras_available, 1)){
     tbvWaitCycle(clk, tbvThreadT::POSEDGE);
   }

   tbvWait(10, TBV_PS);
```

```cpp
    memRasSidebandSignal->mem_ras_type = memRasInterface.mem_ras_type.getValue();
     debugOutput() << "type from ras mon: " << memRasSidebandSignal->mem_ras_type  << endl;

    memRasSidebandSignal->mem_ras_sop = memRasInterface.mem_ras_sop.getValue();
     memRasSidebandSignal->mem_ras_portid = memRasInterface.mem_ras_portid.getValue();
    //get data into vector after reading from mem_ras_data bus ,and create dataPayloadT
    vector<uint8> data(64);// temporary, this must be actual data.

  int dataRead = 0;
//wait while all data is read
  while (dataRead < 48)
    {
         debugOutput() << dec << "@" << tbvGetInt64Time(TBV_NS) << ": Verilog data: MemRas " << mem-
RasInterface.mem_ras_data << endl;
      for (int i = 0; i < 16; i++) {
        data[dataRead] = (memRasInterface.mem_ras_data((127 - 8*i) , (127 - 8*i) - 7).getValue());
        dataRead++;}

       tbvWaitCycle(clk, tbvThreadT::POSEDGE);
       tbvWaitCycle(clk, tbvThreadT::POSEDGE);
       tbvWait(10, TBV_PS);
     }

    debugOutput() << dec << "@" << tbvGetInt64Time(TBV_NS) << ": Verilog data: MemRas " << memRasIn-
terface.mem_ras_data << endl;

  debugOutput() << "data now has size " << dec << data.size() << " dataread is " << dataRead<< hex<< endl;

  for (int i = 0; i < 16; i++)
    {
      data[dataRead] = (memRasInterface.mem_ras_data((127 - 8*i) , (127 - 8*i) - 7).getValue());
      dataRead++;
    }
  data.resize(dataRead);
   dataPayload = new dataPayloadT(data);
   // now you have sideband control and data, bundle them is dataGram and push into mail box.
   dataGram->setDataComP(dataPayload);
   dataGram->setSidebandControlP(memRasSidebandSignal);
    debugOutput() << dec << "MON: data: @ time = " << tbvGetInt64Time(TBV_NS)  << " " << *dataPay-
load<< endl;
    tbvWaitCycle(clk, tbvThreadT::POSEDGE);
   tbvWaitCycle(clk, tbvThreadT::POSEDGE);
   pushIntoMonitorQueue(dataGram);
   tbvThreadT::yield();

}


/*********************************************************************************************/
//                    Mem to Ras Tvm
/*********************************************************************************************/

memRasMonitorTvmT::memRasMonitorTvmT(const char * nameP) :
//  tbvTvmT(nameP),
```

```cpp
reset_l(getFullInterfaceHdlNameP("reset_l") ),
clk(getFullInterfaceHdlNameP("clk") ),
mem_ras_available(getFullInterfaceHdlNameP("mem_ras_available") ),,
mem_ras_data(getFullInterfaceHdlNameP("mem_ras_data") ),
mem_ras_eop(getFullInterfaceHdlNameP("mem_ras_eop") ),
mem_ras_portid(getFullInterfaceHdlNameP("mem_ras_portid") ),
mem_ras_sop(getFullInterfaceHdlNameP("mem_ras_sop") ),
mem_ras_type(getFullInterfaceHdlNameP("mem_ras_type") ),
ras_mem_ctrl_pop(getFullInterfaceHdlNameP("ras_mem_ctrl_pop") ),
ras_mem_data_pop(getFullInterfaceHdlNameP("ras_mem_data_pop") ),
memRasMonitorTask(* this, clk, "Mem Ras Monitor")
{

}

/*****************************************************************************/
memRasMonitorTvmT::~memRasMonitorTvmT(){

}

/*****************************************************************************/
void memRasMonitorTvmT::create(){
  new memRasMonitorTvmT();
}
```

0.2.5 Checker

The common code it shares is common Checker code which has base checker class
and common code for process(), compare() and locate() functions. This is showing
a dummy checker not the actual checker algorithm.

reassemblyChecker.h

```cpp
//-------------------------------------------------------------------//

#ifndef TBV_RASCHECKER_H
#define TBV_RASCHECKER_H
#include "common/topChecker.h"
#include "reassemblyTransform.h"

class reassemblyCheckerT : public topCheckerT {
  public:
    //
    // constructors
    //
    reassemblyCheckerT();
    virtual ~reassemblyCheckerT() { }

    // data access methods

    virtual void processInput(dataGramT * dataGramP);
```

```cpp
    virtual void processOutput(dataGramT * dataGramP);
    virtual bool compare(dataGramT * expectDatacomP, dataGramT * actualDatacomP);
    void body();
    unsigned mismatches;

  private:
    reassemblyTransformT transform;

    // single queues for now, will expand later
    reassemblyInputQueue inputFromMonitor;
    reassemblyOutputQueue outputFromMonitor;
    reassemblyOutputQueue expectedFromChecker;

    tbvSemT cellsToCheck;

    void fakeOutputCellForTesting(dataGramT * dataGramP);

};

#endif
```

reassemblyChecker.cc

```cpp
#include "reassemblyChecker.h"
reassemblyCheckerT::reassemblyCheckerT():
  topCheckerT ("reassembly"),
  mismatches(0)
{}

void reassemblyCheckerT::processInput(dataGramT * dataGramP) {
  debugOutput() << "called reassemblyCheckerT::processInput" << endl;
  inputFromMonitor.push(dataGramP);
}

void reassemblyCheckerT::processOutput(dataGramT * dataGramP) {
  debugOutput() << "called reassemblyCheckerT::processOutput" << endl;
  //for dummy reassembly block
  fakeOutputCellForTesting(dataGramP);
  // else
  // outputFromMonitor.push(dataGramP);
}

void reassemblyCheckerT::fakeOutputCellForTesting(dataGramT * dataGramP) {
  vector<uint8> fakeData = dataGramP->getDataComP()->getDataAsBytes();
  fakeData.resize(52);
  dataPayloadT *fakePayload = new dataPayloadT(fakeData);

    dataGramT *fakeDataGramP = new dataGramT(fakePayload, dataGramP->getSidebandControlP()-
>duplicate());
```

```
    outputFromMonitor.push(fakeDataGramP);
    cellsToCheck.post();
    delete dataGramP;
}


bool reassemblyCheckerT::compare(dataGramT * expectDatacomP, dataGramT * actualDatacomP) {
    dataComT *expect = expectDatacomP->getDataComP();
    dataComT *actual = actualDatacomP->getDataComP();
    if (expect->getDataAsBytes() == actual->getDataAsBytes()) {
        return true;
    } else {
        debugOutput() << "expected: " << *expect << endl;
        debugOutput() << "actual: " << *actual << endl;
        return false;
    }
}

void reassemblyCheckerT::body() {

    cellsToCheck.wait();
    debugOutput() << "called reassemblyCheckerT::body with cell to check" << endl;
    transform.type0(inputFromMonitor,expectedFromChecker);
    dataGramT *actual = outputFromMonitor.front();
    dataGramT *expected = expectedFromChecker.front();
    if (!compare(expected, actual)) {
        mismatches++;
    }
    outputFromMonitor.pop();
    expectedFromChecker.pop();
    delete actual;
    delete expected;
}
```

reassemblyTransform.h

```
#ifndef _REASSEMBLYTRANSFORM_H
#define _REASSEMBLYTRANSFORM_H
#include <stl.h>
#include <queue>
#include "dataGram.h"
#include "dataPayload.h"

typedef vector<unsigned char> byteVector;
class reassemblyInputControl: public sidebandControlT {
private:
  bool sop;
bool eop;
  bool sof;
bool eof;
```

```cpp
public:
    void setStartOfPacket (bool _sop) { sop = _sop; }
    void setEndOfPacket (bool _eop) { eop = _eop; }
    void setStartOfFragment (bool _sof) { sof = _sof; }
    void setEndOfFragment (bool _eof) { eof = _eof; }

    bool getStartOfPacket () { return sop; }
    bool getEndOfPacket () { return eop; }
    bool getStartOfFragment () { return sof; }
    bool getEndOfFragment () { return eof; }

};

class rasInputCell : public dataGramT{
public:
rasInputCell();

};


typedef rasInputCell rasOutputCell;
typedef queue<dataGramT * > reassemblyInputQueue;
typedef queue<dataGramT * > reassemblyOutputQueue;
class reassemblyTransformT {
  public:
    void type0 (reassemblyInputQueue &i, reassemblyOutputQueue &o);
    void type1 (reassemblyInputQueue &i, reassemblyOutputQueue &o);
};

#endif
```

reassemblyTransform.cc

```cpp
#include "reassemblyTransform.h"

rasInputCell::rasInputCell() {

}


void reassemblyTransformT::type0 (reassemblyInputQueue &i, reassemblyOutputQueue &o) {
byteVector inputCell = i.front()->getDataComP()->getDataAsBytes();
rasOutputCell *outputCell = new rasOutputCell();
dataPayloadT *outputPayload = new dataPayloadT (byteVector(inputCell.begin(),inputCell.begin()+52));
outputCell->setDataComP(outputPayload);
o.push(outputCell);
i.pop();
}
```

0.2.6 Stim

Example stimulus test case to drive traffic into RAS interface.

0.2.7 SimpleRasTest.h

```
#ifndef _SIMPLERASTESTCASE_H
#define _SIMPLERASTESTCASE_H

#include "reassemblyTestcase.h"

class simpleRasTest : public reassemblyTestcaseSuiteT {

public:
   virtual void setUp();
   virtual void tearDown();

   CPPUNIT_TEST_SUITE( simpleRasTest );
   CPPUNIT_TEST( testMustPass );
   CPPUNIT_TEST( testProbablyMustPass );
   CPPUNIT_TEST_SUITE_END();

   void testMustPass();
   void testProbablyMustPass();

};

#endif
```

0.2.8 SimpleRasTest.cc

```
#include "simpleRasTest.h"


CPPUNIT_TEST_SUITE_NAMED_REGISTRATION( simpleRasTest, string("reassembly") );

void simpleRasTest::setUp() {
   tbvRandomT::setSeedGenerationAlgorithm(tbvRandomT::RAND48, NULL, 3999, 2);
   // all outputs of integer should show the base
   // ("0" prefix for Oct, "0x" for Hex)
   tbvOut.setf(ios::showbase);

   // explicitly call that which virtual functions will call itself
memRas   reassemblyTestcaseSuiteT::setUp();

}

void simpleRasTest::tearDown() {

   // explicitly call that which virtual functions will call itself
   reassemblyTestcaseSuiteT::tearDown();
```

```
}

void simpleRasTest::testMustPass() {
    tbvOut << "Begin Sample test.........." << endl;
    rasDataProvider->setTotalPackets(10);
    memRasSidebandControlT * rasSidebandControl = new memRasSidebandControlT();
    rasSidebandControl->mem_ras_type = 0;
    rasSidebandControl->mem_ras_qid = 10;
    rasSidebandControl->mem_ras_portid = 1;
    rasDataProvider->setupTraffic(rasSidebandControl);
    tbvOut << "Starting Sample test.........." << endl;

    rasMonitorDispatch->start();
    trafficSchedulerTvm->trafficSchedulerTask.start();
    memRasDriverTvm->memRasDriverTask.start();
    rasChecker->start();

    cout<<"###spawned all tasks"<<endl;
    tbvWait(2000, TBV_NS);

    tbvOut<<"Finished Waiting..." << endl;
    rasMonitorDispatch->stop();
    memRasDriverTvm->memRasDriverTask.stop();
    trafficSchedulerTvm->trafficSchedulerTask.stop();
    rasChecker->stop();

    tbvOut << "Complete Sample test.........." << endl;
    delete rasSidebandControl;

    tbvThreadT::waitDescendants();
    CPPUNIT_ASSERT(rasChecker->mismatches == 0);
}
void simpleRasTest::testProbablyMustPass() {
    testMustPass();
}
```

0.3   Example for standard interfaces:

In this appendix section is given the code implemented for the standard transactors
in testBuilder. The flow and other details for the transactors are in respective sections. The code here is to show the detailed implementation of the interfaces and is
not dependent on common infrastructure code. Only common classes are the cell
and packet classes defined.

0.3.1 Code Example for PL3Tx Standard interface:

pl3TxGen.h

```
#ifndef _pl3TXGEN_H
```

```cpp
#define _pl3TXGEN_H

#include "TestBuilder.h"
#include "tbvCell.h"
#include "tbvPkt.h"

class pl3TxGenTvmT;

// ************************************
// a task pl3TxGen
// ************************************

class pl3TxGenTaskT : public tbvTaskTypeSafeT<tbvCellT> {
  public:
    //parent tvm, make it easy to access in task body
    pl3TxGenTvmT& pl3TGen;
    tbvSmartUnsignedT randBit; // To generate random bits

    //constructor, destructor
    pl3TxGenTaskT ( pl3TxGenTvmT& );
    virtual ~pl3TxGenTaskT();

    //body function for task functionality
    virtual void body ( tbvCellT *arg );
};


// ************************************
// a task pl3TxQueue
// ************************************

class pl3TxQueueTaskT : public tbvTaskTypeSafeT<tbvCellT> {
  private:
    pl3TxGenTvmT& pl3TGen;
    pl3TxGenTaskT& mainTxTask;

  public:
    pl3TxQueueTaskT (pl3TxGenTvmT& pl3TGen, pl3TxGenTaskT& mainTxTask);
    virtual ~pl3TxQueueTaskT();

    //body function for task functionality
    virtual void body ( tbvCellT *arg );
};


// ************************************
// a task pl3TxFull
// ************************************

class pl3TxFullTaskT : public tbvTaskT {
  private:
    pl3TxGenTvmT& pl3TGen;
    pl3TxGenTaskT& mainTxTask;

  public:
    pl3TxFullTaskT (pl3TxGenTvmT& pl3TGen, pl3TxGenTaskT& mainTxTask);
    virtual ~pl3TxFullTaskT();

    //body function for task functionality
    virtual void body (tbvSmartDataT *);
};
```

```cpp
// ***************************************
// a task pl3TxStat
// ***************************************

class pl3TxStatTaskT : public tbvTaskTypeSafeT<tbvCellT> {
  public:
    //parent tvm, make it easy to access in task body
    pl3TxGenTvmT& pl3TGen;

    //constructor, destructor
    pl3TxStatTaskT ( pl3TxGenTvmT& );
    virtual ~pl3TxStatTaskT();

    //body function for task functionality
    virtual void body ( tbvCellT *arg );
};


// ***************************************
// a task runStat
// ***************************************

class runStatTaskT : public tbvTaskT
{
  public:
    //parent tvm, make it easy to access in task body
    pl3TxGenTvmT& pl3TGen;

    //constructor, destructor
    runStatTaskT ( pl3TxGenTvmT& );
    virtual ~runStatTaskT();

    //body function for task functionality
    virtual void body ( tbvSmartDataT * );
};


// ***************************************
// a TVM pl3TxGen
// ***************************************


class pl3TxGenTvmT : public tbvTvmT
{

  public:
    //constructor, destructor
    pl3TxGenTvmT ();
    virtual ~pl3TxGenTvmT();

    //hdl interface signals
tbvSignalHdlT   pp3tfclk;
tbvSignalHdlT   rst_l;
tbvSignalHdlT   pp3tenb_n;
tbvSignalHdlT   pp3tsx;
tbvSignalHdlT   pp3tadr;
tbvSignalHdlT   pp3tdat;
tbvSignalHdlT   pp3tsop;
tbvSignalHdlT   pp3teop;
tbvSignalHdlT   pp3tmod;
tbvSignalHdlT   pp3tprty;
```

```cpp
    tbvSignalHdlT   pp3terr;
    tbvSignalHdlT   pp3ptpa;
    tbvSignalHdlT   pp3stpa;


  // The posPhyIf Mutex
  tbvMutexT pl3TxGenMutex;
  tbvMutexT pl3TxQueueMutex;
  tbvMutexT pl3TxFullMutex;
  tbvMutexT pl3TxStatMutex;
  tbvMutexT fullStatQueueMutex;

  // the recording fiber is needed to avoid using the parent task fiber
  tbvFiberT pl3TxGenFiber;
  tbvFiberT pl3TxQueueFiber;
  tbvFiberT pl3TxFullFiber;

  //basic tasks
  pl3TxGenTaskT pl3TxGen;
  pl3TxQueueTaskT pl3TxQueue;
  pl3TxFullTaskT pl3TxFull;
  pl3TxStatTaskT pl3TxStat;
  runStatTaskT runStat;
  bool running;
  void start(){running = true;}
  void stop();

  unsigned int port_stat[64];
  unsigned int port_cnt;
  unsigned int prtySense;
  unsigned int badParity;
  unsigned int badMode;
  unsigned int curr_port;
  list<int> port_list;
  vector<int> port_index;

  void setPort_cnt(unsigned int k){port_cnt = k;}
  void setPrtySense(unsigned int k){prtySense = k;}
  void setBadParity(unsigned int k){badParity = k;}
  void setBadMode(unsigned int k){badMode = k;}
  void setPort_list(list<int> & k){port_list = k;}

  unsigned int statFull;
  void setStatFull(unsigned int k){statFull = k;}

  vector<tbvCellT *> fullStatQueue;

  // Event Expressions
  tbvEventExprT posedge_clk;
  tbvEventExprT not_reset;

  // This function is called from $tbv_tvm_connect, to Create a TVM object.
  static void create ( );


};


#endif

/**************************************************/
```

pl3TxGen.cc

```cpp
#include "pl3TxGen.h"
#include "TestBuilder.h"


// ******************************************************************
//                    pl3TxGen Task
// ******************************************************************

//
// Tvm pl3TGen Task pl3TxGen Constructor & Destructor
//
pl3TxGenTaskT::pl3TxGenTaskT ( pl3TxGenTvmT & tvm ) :
  tbvTaskTypeSafeT<tbvCellT> ( &tvm, "pl3TxGen" ),

  // TVM reference
  pl3TGen ( tvm )
{
 // Disable the mutex built into the fiber and turn off automatic
 // transaction recording
 setSynchronization(FALSE, FALSE);
}

pl3TxGenTaskT::~pl3TxGenTaskT () {}

//
// Tvm pl3TGen Task pl3TxGen body method
//
void pl3TxGenTaskT::body ( tbvCellT *pl3TxGenArgP )
{
  tbvOut << pl3TGen.getInstanceNameP() << " is called @ ";
  tbvOut << tbvGetInt64Time(TBV_NS) << " ns\n";

  //
  // Do one transaction at a time
  //
  pl3TGen.pl3TxGenMutex.lock();

  vector<uint32> data = pl3TxGenArgP->getDataBy32();
  bool prty = 0;

  tbvSmartUnsignedT badPrtyRand;
  badPrtyRand.keepOnly(1,100);

  int beginTime = tbvGetInt64Time(TBV_NS);
  int loopTime = beginTime;

  while (pl3TxGenArgP->getValid() && (pl3TGen.port_stat[(pl3TxGenArgP->getInputPort().getUn-
signedValue())] == 0)) {
    loopTime = tbvGetInt64Time(TBV_NS);
    tbvWaitCycle(pl3TGen.pp3tfclk, tbvThreadT::POSEDGE);
    tbvWait(0.5, TBV_NS); // Hold time
    // 64 ports * 160 ns (Oc48) per port = 10240 = max port calendar
    if (loopTime - beginTime > 20000)
      {
tbvExceptionT::setUserExceptionTypeString("PHY INPUT STATUS 0 KEPT TOO LONG");
tbvExceptionT::reportUserException("PHY INPUT STATUS 0 KEPT TOO LONG",
  tbvExceptionT::ERROR,
```

```
    "Waited 20000 ns for port_stat = 1 on port %d", pl3TxGenArgP->getInputPort().getUnsigned-
Value() );

tbvExit();
    }
  }

  if (loopTime > beginTime) {
    tbvExceptionT::setUserExceptionTypeString("PHY CORE BACKPRESSURE");
    tbvExceptionT::reportUserException("PHY CORE BACKPRESSURE",
                        tbvExceptionT::WARNING,
                        "Backpressure on port %d for %d ns (%d to %d ns)", pl3TxGenArgP->get-
InputPort().getUnsignedValue(), (loopTime - beginTime), beginTime, loopTime );
  }

  tbvSignalT dataSig(31,0); // Used because easier to calculate parity

  //Start transaction recording
  if (pl3TxGenArgP->getValid())
    pl3TGen.pl3TxGenFiber.beginTransactionH("Valid Cell");
  else
    pl3TGen.pl3TxGenFiber.beginTransactionH("Invalid Cell");

  // If port has changed from last time and cell is valid, send new port
  dataSig = pl3TxGenArgP->getInputPort().getUnsignedValue();
  if ((dataSig.getValue() != pl3TGen.curr_port) && pl3TxGenArgP->getValid()) {
    pl3TGen.curr_port = dataSig.getValue();
    pl3TGen.pp3tenb_n = 1;
    pl3TGen.pp3tsx = 1;
    randBit.randomize();
    dataSig = pl3TxGenArgP->getInputPort().getUnsignedValue() + ((randBit % 16777216) << 8); //
Only lsb 8-bits should be considered
    pl3TGen.pp3tdat = dataSig.getValue();

    prty = dataSig.reductionXor();;
    badPrtyRand.randomize();
    if (badPrtyRand <= pl3TGen.badParity)
      prty = prty ^ 1;
    if (pl3TGen.prtySense)
      pl3TGen.pp3tprty = prty ^ 1;
    else
      pl3TGen.pp3tprty = prty;

    // Randomize don't-care signals
    randBit.randomize();
    pl3TGen.pp3tmod = randBit % 4;
    randBit.randomize();
    pl3TGen.pp3tsop = randBit % 2;
    randBit.randomize();
    pl3TGen.pp3teop = randBit % 2;
    randBit.randomize();
    pl3TGen.pp3terr = randBit % 2;

    tbvWaitCycle(pl3TGen.pp3tfclk, tbvThreadT::POSEDGE);
    tbvWait(0.5, TBV_NS); // Hold time
  }

  // Send data
  for (unsigned int i = 0; i < data.size(); i++) {

    pl3TGen.pp3tdat = data[i];
```

```cpp
    // Calculate parity
    dataSig = data[i];
    if (pl3pl3TxGenArgP->getValid()) {
      prty = dataSig.reductionXor();
      badPrtyRand.randomize();
      if (badPrtyRand <= pl3TGen.badParity)
prty = prty ^ 1;
      if (pl3TGen.prtySense)
pl3TGen.pp3tprty = prty ^ 1;
      else
pl3TGen.pp3tprty = prty;
    }
    else { // parity ignored when tenb, tsx disabled so randomize it
      randBit.randomize();
      pl3TGen.pp3tprty = randBit % 2;
    }

    // Start of burst
    if (i == 0) {
      pl3TGen.pp3tenb_n = (1 - pl3TxGenArgP->getValid());
      pl3TGen.pp3tsop = pl3TxGenArgP->getSOP();
      if (data.size() >= 2) { // Not end of burst
pl3TGen.pp3teop = 0;
pl3TGen.pp3terr = 0;
pl3TGen.pp3tmod = 0;
      }
    }

    if (i == 1)  pl3TGen.pp3tsop = 0;

    if (i == (data.size()-1)) {
      pl3TGen.pp3teop = pl3TxGenArgP->getEOP();
      pl3TGen.pp3tmod = (pl3TGen.badMode) ? ((4 - (pl3TxGenArgP->getSize()%4)) % 4) :
((pl3TxGenArgP->getEOP()) ? ((4 - (pl3TxGenArgP->getSize()%4)) % 4) : 0);
      // terr asserted only when teop is asserted
      if (pl3TxGenArgP->getEOP())
pl3TGen.pp3terr = pl3TxGenArgP->getBad();
      else
pl3TGen.pp3terr = 0;
    }

    // tsx is ignored when tenb_n is 0, so randomize it
    if (pl3TxGenArgP->getValid()) {
      randBit.randomize();
      pl3TGen.pp3tsx = randBit % 2;
    }
    else
      pl3TGen.pp3tsx = 0;

    tbvWaitCycle(pl3TGen.pp3tfclk, tbvThreadT::POSEDGE);
    tbvWait(0.5, TBV_NS); // Hold time
  }

  // In case another cell is not immediately generated after this one.
  pl3TGen.pp3tenb_n = 1;
  pl3TGen.pp3tsx = 0;
  if (pl3TxGenArgP->getValid() == 1)
    pl3TGen.port_stat[pl3TxGenArgP->getInputPort().getUnsignedValue()] = 0;

  //Record the values in the argument block
  pl3TGen.pl3TxGenFiber.recordAttribute(*pl3TxGenArgP, "Cell");
```

```cpp
//Finish Transaction Recording
pl3TGen.pl3TxGenFiber.endTransaction();

// release the semaphore
pl3TGen.pl3TxGenMutex.unlock();

tbvOut << pl3TGen.getInstanceNameP() << " done @ " << tbvGetInt64Time(TBV_NS) << " ns\n";

}


// ****************************************************************
//                    pl3TxQueue Task
// ****************************************************************

//
// Tvm pl3TGen Task pl3TxQueue Constructor & Destructor
//
pl3TxQueueTaskT::pl3TxQueueTaskT (pl3TxGenTvmT& pl3TGen, pl3TxGenTaskT& mainTxTask )
:
   tbvTaskTypeSafeT<tbvCellT> ( &pl3TGen, "pl3TxQueue" ),
   pl3TGen(pl3TGen),
   mainTxTask(mainTxTask)

{
 // Disable the mutex built into the fiber and turn off automatic
 // transaction recording
 setSynchronization(FALSE, FALSE);
}

pl3TxQueueTaskT::~pl3TxQueueTaskT () {}

//
// Tvm pl3TGen Task pl3TxQueue body method
//
void pl3TxQueueTaskT::body ( tbvCellT *pl3TxGenArgP )
{
 //
 // Do one transaction at a time
 //
 pl3TGen.pl3TxQueueMutex.lock();
 pl3TGen.fullStatQueueMutex.lock();

 //
 // wait for Reset to end (if it is asserted)
 //

 if (tbv4StateNotEqual(pl3TGen.rst_l, 1))
   tbvWait( pl3TGen.not_reset && pl3TGen.posedge_clk , tbvEventExprT::RWSYNC );

 // Push the current transfer request into the FIFO
 pl3TGen.fullStatQueue.push_back(pl3TxGenArgP);

 tbvCellT *queueArg = NULL;

 unsigned int i = 0;
 unsigned int main_task_run = 0;
 unsigned int flag_stat[64];

 for (unsigned int i = 0; i < 64; i++)
    flag_stat[i] = 0;
```

```cpp
  // Go through all FIFO entries
  while (i < pl3TGen.fullStatQueue.size())
   if (pl3TGen.fullStatQueue.size() > 0)
    {
      queueArg = pl3TGen.fullStatQueue[i];
      flag_stat[queueArg->getInputPort().getUnsignedValue()] = flag_stat[queueArg->getInput-
Port().getUnsignedValue()] | (((pl3TGen.port_stat[(queueArg->getInputPort().getUnsignedValue())
% 64] == 0)) ? 1 : 0);


      // Send the current transfer request if the status (per port) allows or if the data is invalid
      if (((pl3TGen.port_stat[(queueArg->getInputPort().getUnsignedValue()) % 64] != 0) &&
(flag_stat[queueArg->getInputPort().getUnsignedValue()] == 0)) || (queueArg->getValid() == 0))
       {
         if (pl3TGen.fullStatQueue.size() > 0)
          {
            mainTxTask.run(queueArg);
            main_task_run = 1;
          }
         if (pl3TGen.fullStatQueue.size() > 0)
          {
            // Shift up all entirs in the FIFO and erase the last one
            for (unsigned int j = i; j < pl3TGen.fullStatQueue.size() - 1; j++)
              {
                 pl3TGen.fullStatQueue[j] = pl3TGen.fullStatQueue[j+1];
              }
            pl3TGen.fullStatQueue.pop_back();
            i = i - 1;
          }
       }
     i = i + 1;
    }

  // Send invalid cell if nothing can be popped from the FIFO
  tbvCellT invalid_cell;
  vector<uint8> pkt(64);
  tbvControlDataT control;
  control.fid.input_port = 0;
  control.valid = 0;
  control.sop = 0;
  control.eop = 0;

  for (int j = 0; j < 64; j++)
     pkt[j] = j;
  invalid_cell.setPayload(pkt);
  invalid_cell.setControlData(control);

  if (main_task_run == 0)
   {
     mainTxTask.run(&invalid_cell);
     //tbvOut << "INVALID CELL" << endl;
   }


  pl3TGen.fullStatQueueMutex.unlock();
  pl3TGen.pl3TxQueueMutex.unlock();


}


// ****************************************************************
//                     pl3TxFull Task
```

```cpp
// ************************************************************

//
// Tvm pl3TGen Task pl3TxFull Constructor & Destructor
//
pl3TxFullTaskT::pl3TxFullTaskT (pl3TxGenTvmT& pl3TGen, pl3TxGenTaskT& mainTxTask ) :
   tbvTaskT ( &pl3TGen, "pl3TxFull" ),
   pl3TGen(pl3TGen),
   mainTxTask(mainTxTask)


{
  // Disable the mutex built into the fiber and turn off automatic
  // transaction recording
  setSynchronization(FALSE, FALSE);
}

pl3TxFullTaskT::~pl3TxFullTaskT () {}

//
// Tvm pl3TGen Task pl3TxFull body method
//
void pl3TxFullTaskT::body (tbvSmartDataT *pl3TxGenArgP)
{
  //
  // Do one transaction at a time
  //
  pl3TGen.pl3TxFullMutex.lock();
  pl3TGen.fullStatQueueMutex.lock();

  //
  // wait for Reset to end (if it is asserted)
  //

  if (tbv4StateNotEqual(pl3TGen.rst_l, 1))
    tbvWait( pl3TGen.not_reset && pl3TGen.posedge_clk , tbvEventExprT::RWSYNC );

  tbvCellT *queueArg = NULL;

  unsigned int i = 0;
  unsigned int stat_flag[64];

  for (unsigned int i = 0; i < 64; i++)
    stat_flag[i] = 0;

  // Go through all FIFO entries
  while (i < pl3TGen.fullStatQueue.size())
   if (pl3TGen.fullStatQueue.size() > 0)
    {
      queueArg = pl3TGen.fullStatQueue[i];
      stat_flag[queueArg->getInputPort().getUnsignedValue()] = stat_flag[queueArg->getInput-
Port().getUnsignedValue()] | (((pl3TGen.port_stat[(queueArg->getInputPort().getUnsignedValue())
% 64] == 0)) ? 1 : 0);


      // Send the current transfer request if the status (per port) allows or if the data is invalid
      if ((((pl3TGen.port_stat[(queueArg->getInputPort().getUnsignedValue()) % 64] != 0) &&
(stat_flag[queueArg->getInputPort().getUnsignedValue()] == 0)) || (queueArg->getValid() == 0))
        {
          if (pl3TGen.fullStatQueue.size() > 0)
            {
              mainTxTask.run(queueArg);
            }
```

```cpp
      if (pl3TGen.fullStatQueue.size() > 0)
        {
          // Shift up all entirs in the FIFO and erase the last one (erase did not work)
          for (unsigned int j = i; j < pl3TGen.fullStatQueue.size() - 1; j++)
            {
              pl3TGen.fullStatQueue[j] = pl3TGen.fullStatQueue[j+1];
            }
          pl3TGen.fullStatQueue.pop_back();
          i = i - 1;
        }
    }
  i = i + 1;
}


  pl3TGen.fullStatQueueMutex.unlock();
  pl3TGen.pl3TxFullMutex.unlock();

}


// ******************************************************************
//                     pl3TxStat Task
// ******************************************************************

//
// Tvm pl3TGen Task pl3TxStat Constructor & Destructor
//
pl3TxStatTaskT::pl3TxStatTaskT ( pl3TxGenTvmT & tvm ) :
   tbvTaskTypeSafeT<tbvCellT> ( &tvm, "pl3TxStat" ),

   // TVM reference
   pl3TGen ( tvm )
{
  // Disable the mutex built into the fiber and turn off automatic
  // transaction recording
  setSynchronization(FALSE, FALSE);
}

pl3TxStatTaskT::~pl3TxStatTaskT () {}

//
// Tvm pl3TGen Task pl3TxStat body method
//
void pl3TxStatTaskT::body ( tbvCellT *pl3TxGenArgP )
{
  //
  // Do one transaction at a time
  //

  pl3TGen.pl3TxStatMutex.lock();

  //
  // wait for Reset to end (if it is asserted)
  //

  if (tbv4StateNotEqual(pl3TGen.rst_l, 1))
    tbvWait( pl3TGen.not_reset && pl3TGen.posedge_clk , tbvEventExprT::RWSYNC );

  unsigned int old_port_stat;
  unsigned int port_count = pl3TGen.port_cnt;
  //Collect the list of ports
```

```cpp
for(list<int>::iterator i = pl3TGen.port_list.begin(); i != pl3TGen.port_list.end(); i++ )
  pl3TGen.port_index.push_back(*i);

unsigned int j = 0;
for(list<int>::iterator i = pl3TGen.port_list.begin(); i != pl3TGen.port_list.end(); i++ )
  {
    // Get the status of the ports two clk cycles after the polling of the addresses
    unsigned int adr_cnt =  *i;
    unsigned int port = port_count + j - 3;
    pl3TGen.pp3tadr = adr_cnt % 64;
    tbvWait(0.5, TBV_NS);
    old_port_stat = pl3TGen.port_stat[pl3TGen.port_index[port % port_count]];
    pl3TGen.port_stat[pl3TGen.port_index[port % port_count]] = ((pl3TGen.pp3ptpa.getValue())) %
2;
    //tbvOut << dec << "@ " << tbvGetInt64Time(TBV_NS)  << "Port_Status " <<
pl3TGen.port_stat[pl3TGen.port_index[port % port_count]] << "  ADDR " << (port % port_count) <<
" Port_index " << pl3TGen.port_index[port % port_count] << " Port_adr_cnt " << port << "
Old_Port_Status " << old_port_stat << " PORT_ADDR : " << adr_cnt << endl;
    //If the status (per port) changes from 0 to 1 and the queue is not empty -> spawn the
pl3TxFullTask one more time
    if (((old_port_stat == 0) && (pl3TGen.port_stat[pl3TGen.port_index[port % port_count]] == 1))
&& (pl3TGen.fullStatQueue.size() > 0))
      {
        pl3TGen.pl3TxFull.spawn();
        //tbvOut << "SPAWN at : " << tbvGetInt64Time(TBV_NS) << " ns"  << endl;
      }

    j = j + 1;
    tbvWaitCycle(pl3TGen.pp3tfclk, tbvThreadT::POSEDGE);

  }

pl3TGen.port_index.clear();

// release the semaphore
pl3TGen.pl3TxStatMutex.unlock();

}


// ****************************************************************
// a task runStat
// ****************************************************************

//
// Tvm pl3TGen Task runStat Constructor & Destructor
//
runStatTaskT::runStatTaskT ( pl3TxGenTvmT & tvm ) :
  tbvTaskT ( &tvm, "runStatTask" ),


  // TVM reference
  pl3TGen ( tvm )
{
  // Disable the mutex built into the fiber and turn off automatic
  // transaction recording
  setSynchronization(FALSE, FALSE);
}

runStatTaskT::~runStatTaskT () {}
void runStatTaskT::body ( tbvSmartDataT *pl3TxGenArgP )
{
```

```cpp
//for (unsigned int i = 0; i < 64; i++)
 //pl3TGen.port_stat[i] = 0;

 while(pl3TGen.running && !pl3TGen.port_list.empty())
  {
   pl3TGen.pl3TxStat.run();
   tbvThreadT::yield();
  }
}


//****************************************
// stop function
//****************************************
void pl3TxGenTvmT::stop()
 {
   while (fullStatQueue.size() > 0)
    {
      tbvWaitCycle(pp3tfclk, tbvThreadT::POSEDGE);
    }
   running = false;
 }




// ******************************************************************
//                  pl3TGen TVM
// ******************************************************************

//
// TVM pl3TGen Constructor & Destructor
//
pl3TxGenTvmT::pl3TxGenTvmT ( ) :
 tbvTvmT (),

 // Initialize the TVM Signals
     pp3tfclk (getFullInterfaceHdlNameP("pp3tfclk") ),
     rst_l (getFullInterfaceHdlNameP("rst_l") ),
     pp3tenb_n (getFullInterfaceHdlNameP("pp3tenb_n_o") ),
     pp3tsx (getFullInterfaceHdlNameP("pp3tsx_o") ),
     pp3tadr (getFullInterfaceHdlNameP("pp3tadr_o") ),
     pp3tdat (getFullInterfaceHdlNameP("pp3tdat_o") ),
     pp3tsop (getFullInterfaceHdlNameP("pp3tsop_o") ),
     pp3teop (getFullInterfaceHdlNameP("pp3teop_o") ),
     pp3tmod (getFullInterfaceHdlNameP("pp3tmod_o") ),
     pp3tprty (getFullInterfaceHdlNameP("pp3tprty_o") ),
     pp3terr (getFullInterfaceHdlNameP("pp3terr_o") ),
     pp3ptpa (getFullInterfaceHdlNameP("pp3ptpa") ),
     pp3stpa (getFullInterfaceHdlNameP("pp3stpa") ),


 // Initialize the Mutex
 pl3TxGenMutex( "pl3TxGenMutex" ),
 pl3TxQueueMutex( "pl3TxQueueMutex" ),
 pl3TxFullMutex( "pl3TxFullMutex" ),
 pl3TxStatMutex( "pl3TxStatMutex" ),
 fullStatQueueMutex( "fullStatQueueMutex" ),

 pl3TxGenFiber( this, "pl3TxGenFiber"),
 pl3TxQueueFiber( this, "pl3TxQueueFiber"),
 pl3TxFullFiber( this, "pl3TxFullFiber"),
```

```
    // Create the task objects
    pl3TxGen(* this),
    pl3TxQueue(* this, pl3TxGen),
    pl3TxFull(* this, pl3TxGen),
    pl3TxStat(* thi),
    runStat(* this),
    running(false),
    port_cnt(64),
    prtySense(0),
    badParity(0),
    badMode(0),
    curr_port(0xFF), // unused
    statFull(0),

    // Create the Event Expressions
    posedge_clk ( pp3tfclk, tbvThreadT::POSEDGE ),
    not_reset ( rst_l , 1 )

{
 //add any construtor code needed for tvm pl3TGen
 pl3TxGenFiber.setRecording(TRUE);
}

//destructor for tvm pl3TGen
pl3TxGenTvmT::~pl3TxGenTvmT()
{
  //add any clean up code needed for tvm pl3TGen

}

//
// Method called from $tbv_tvm_connect, to create the TVM object.
//
void pl3TxGenTvmT::create ( )
{
  new pl3TxGenTvmT ( );
};


/********************************************************/
```

0.3.2 Code Example for Standard PL3Rx interface

PL3RxGen.h

```
#include "TestBuilder.h"
#include "tbvCell.h"
#include "tbvPkt.h"

class pl3RxGenTvmT;

// ***************************************
// a task pp3RxGen
// ***************************************

class pp3RxGenTaskT : public tbvTaskTypeSafeT<tbvCellT> {
```

```cpp
public:
  //parent tvm, make it easy to access in task body
  pl3RxGenTvmT& pp3RGen;
  tbvSmartUnsignedT randBit; // To generate random bits

  //constructor, destructor
  pp3RxGenTaskT ( pl3RxGenTvmT& );
  virtual ~pp3RxGenTaskT();

  //body function for task functionality
  virtual void body ( tbvCellT *arg );
};


// ************************************
// a TVM pl3RxGen
// ************************************


class pl3RxGenTvmT : public tbvTvmT
{

  public:
    //constructor, destructor
    pl3RxGenTvmT ();
    virtual ~pl3RxGenTvmT();

    //hdl interface signals
  tbvSignalHdlT  pp3rfclk;
  tbvSignalHdlT  rst_l;
  tbvSignalHdlT  pp3renb_n;
  tbvSignalHdlT  pp3rsx;
  tbvSignalHdlT  pp3rsop;
  tbvSignalHdlT  pp3reop;
  tbvSignalHdlT  pp3rdat;
  tbvSignalHdlT  pp3rmod;
  tbvSignalHdlT  pp3rprty;
  tbvSignalHdlT  pp3rerr;
  tbvSignalHdlT  pp3rval;


    // The pl3If Mutex
    tbvMutexT pp3RxGenMutex;

    // the recording fiber is needed to avoid using the parent task fiber
    tbvFiberT pp3RxGenFiber;

    //basic tasks
    pp3RxGenTaskT pp3RxGen;

    // Event Expressions
    tbvEventExprT posedge_clk;
    tbvEventExprT pp3renb_n_0;

    unsigned int prtySense;
    void setPrtySense(unsigned int k){prtySense = k;}

    unsigned int badParity;
    void setBadParity(unsigned int k){badParity = k;}

    unsigned int badMode;
    void setBadMode(unsigned int k){badMode = k;}
```

```cpp
    unsigned int curr_port;

    // This function is called from $tbv_tvm_connect, to Create a TVM object.
    static void create ( );


};
/*********************************************************************/
```

pl3RxGen.cc

```cpp
#include "pl3RxGen.h"
#include "TestBuilder.h"



// ****************************************************************
//                  pp3RxGen Task
// ****************************************************************


//
// Tvm pp3RGen Task pp3RxGen Constructor & Destructor
//
pp3RxGenTaskT::pp3RxGenTaskT ( pl3RxGenTvmT & tvm ) :
    tbvTaskTypeSafeT<tbvCellT> ( &tvm, "pp3RxGen" ),

    // TVM reference
    pp3RGen ( tvm )
{
  // Disable the mutex built into the fiber and turn off automatic
  // transaction recording
  setSynchronization(FALSE, FALSE);
}

pp3RxGenTaskT::~pp3RxGenTaskT () {}

//
// Tvm pp3RGen Task pp3RxGen body method
//
void pp3RxGenTaskT::body ( tbvCellT *pp3RxGenArgP )
{
  tbvOut << pp3RGen.getInstanceNameP() << " is called @ ";
  tbvOut << tbvGetInt64Time(TBV_NS) << " ns\n";

  //
  // Lock mutex to do  one transaction at a time
  //
  pp3RGen.pp3RxGenMutex.lock();

  vector<uint32> data = pp3RxGenArgP->getDataBy32();
  bool prty = 0;

  tbvSmartUnsignedT badPrtyRand;
  badPrtyRand.keepOnly(1,100);

  // Signals change only on rising edge of clk
  if (pp3RGen.pp3rfclk == 0) {
    tbvWaitCycle(pp3RGen.pp3rfclk, tbvThreadT::POSEDGE);
    tbvWait(1.5, TBV_NS); // Minimum hold time
  }

  // pl3 Enable signal controls the sender
  if (pp3RGen.pp3renb_n != 0)
```

```
    pp3RGen.handleBackpressure();

  tbvSignalT dataSig(31,0); // Used because easier to calculate parity

  if (pp3RxGenArgP->getValid())
    pp3RGen.pp3RxGenFiber.beginTransactionH("Valid Cell");
  else
    pp3RGen.pp3RxGenFiber.beginTransactionH("Invalid Cell");

  // If port has changed from last time, send new port
  dataSig = pp3RxGenArgP->getInputPort().getUnsignedValue();
  if (dataSig.getValue() != pp3RGen.curr_port) {
    pp3RGen.curr_port = dataSig.getValue();
    pp3RGen.pp3rsx = 1;
    pp3RGen.pp3rval = 0;
    randBit.randomize();
    dataSig = (pp3RxGenArgP->getInputPort().getUnsignedValue()) + ((randBit % 16777216) << 8); /
/ Only lsb 8-bits should be considered
    pp3RGen.pp3rdat = dataSig.getValue();

    prty = dataSig.reductionXor();
    badPrtyRand.randomize();
    if (badPrtyRand <= pp3RGen.badParity)
      prty = prty ^ 1;
    if (pp3RGen.prtySense)
      pp3RGen.pp3rprty = prty ^ 1;
    else
      pp3RGen.pp3rprty = prty;

    // Randomize don't-care signals
    randBit.randomize();
    pp3RGen.pp3rmod = randBit % 4;
    randBit.randomize();
    pp3RGen.pp3rsop = randBit % 2;
    randBit.randomize();
    pp3RGen.pp3reop = randBit % 2;
    randBit.randomize();
    pp3RGen.pp3rerr = randBit % 2;

    tbvWaitCycle(pp3RGen.pp3rfclk, tbvThreadT::POSEDGE);
    tbvWait(1.5, TBV_NS); // Minimum hold time
  }

  // Send data
  for(unsigned i = 0; i < data.size(); i++) {
    if (tbv4StateNotEqual(pp3RGen.pp3renb_n, 0)) {
      pp3RGen.handleBackpressure();
    }

    pp3RGen.pp3rdat = data[i];

    dataSig = data[i];
    if (pp3RxGenArgP->getValid()) {
      prty = dataSig.reductionXor();
      badPrtyRand.randomize();
      if (badPrtyRand <= pp3RGen.badParity)
prty = prty ^ 1;
      if (pp3RGen.prtySense)
pp3RGen.pp3rprty = prty ^ 1;
      else
pp3RGen.pp3rprty = prty;
    }
```

```
      else { // parity ignored when rval, rsx disabled so randomize it
        randBit.randomize();
        pp3RGen.pp3rprty = randBit % 2;
      }

      // Start of burst
      if (i == 0) {
        pp3RGen.pp3rval = pp3RxGenArgP->getValid();
        pp3RGen.pp3rsop = pp3RxGenArgP->getSOP();
        if (data.size() >= 2) { // Not end of burst
  pp3RGen.pp3reop = 0;
  pp3RGen.pp3rerr = 0;
  pp3RGen.pp3rmod = 0;
        }
      }

      if (i == 1)  pp3RGen.pp3rsop = 0;

      if (i == (data.size()-1)) {
        pp3RGen.pp3reop = pp3RxGenArgP->getEOP();
        pp3RGen.pp3rmod = (pp3RGen.badMode) ? ((4 - (pp3RxGenArgP->getSize()%4)) % 4) :
  ((pp3RxGenArgP->getEOP()) ? ((4 - (pp3RxGenArgP->getSize()%4)) % 4) : 0);
        // rerr asserted only when reop is asserted
        if(pp3RxGenArgP->getEOP())
  pp3RGen.pp3rerr = pp3RxGenArgP->getBad();
      }

      // Randomize rsx when rval = 1
      if (pp3RxGenArgP->getValid()) {
        randBit.randomize();
        pp3RGen.pp3rsx = randBit % 2;
      }
      else
        pp3RGen.pp3rsx = 0;

      tbvWaitCycle(pp3RGen.pp3rfclk, tbvThreadT::POSEDGE);
      tbvWait(1.5, TBV_NS); // Minimum hold time
    }

    // In case this was the last cell, then, we need to get rval and rsx to 0.
    // However, if renb_n != 0, we first have to wait until it is 1 before
    // we do so because rval and rsx have to stay the same while renb_n != 0
    if (pp3RGen.pp3renb_n != 0)
      pp3RGen.handleBackpressure();
    pp3RGen.pp3rval = 0;
    pp3RGen.pp3rsx = 0; // If data.size() was zero

    // Record the values in the argument block
    pp3RGen.pp3RxGenFiber.recordAttribute(*pp3RxGenArgP, "Cell");

    // Finish Transaction Recording
    pp3RGen.pp3RxGenFiber.endTransaction();

    // release the semaphore
    pp3RGen.pp3RxGenMutex.unlock();

    tbvOut << pp3RGen.getInstanceNameP() << " done @ " << tbvGetInt64Time(TBV_NS) << " ns\n";

}

// ****************************************************************
//                  pp3RGen TVM
```

```cpp
// *****************************************************************
//
// TVM pp3RGen Constructor & Destructor
//
pl3RxGenTvmT::pl3RxGenTvmT ( ) :
  tbvTvmT (),

    // Initialize the TVM Signals with the actual verilog signals
        pp3rfclk (getFullInterfaceHdlNameP("pp3rfclk") ),
        rst_l (getFullInterfaceHdlNameP("rst_l") ),
        pp3renb_n (getFullInterfaceHdlNameP("pp3renb_n_r") ),
        pp3rsx (getFullInterfaceHdlNameP("pp3rsx_o") ),
        pp3rsop (getFullInterfaceHdlNameP("pp3rsop_o") ),
        pp3reop (getFullInterfaceHdlNameP("pp3reop_o") ),
        pp3rdat (getFullInterfaceHdlNameP("pp3rdat_o") ),
        pp3rmod (getFullInterfaceHdlNameP("pp3rmod_o") ),
        pp3rprty (getFullInterfaceHdlNameP("pp3rprty_o") ),
        pp3rerr (getFullInterfaceHdlNameP("pp3rerr_o") ),
        pp3rval (getFullInterfaceHdlNameP("pp3rval_o") ),


    // Initialize the Mutex
    pp3RxGenMutex( "pp3RxGenMutex" ),
    pp3RxGenFiber( this, "pp3RxGenFiber"),

    // Create the task objects
    pp3RxGen(* this),

    // Create the Event Expressions
    posedge_clk ( pp3rfclk, tbvThreadT::POSEDGE ),
    pp3renb_n_0(posedge_clk, !pp3renb_n()),

    prtySense(0),
    badParity(0),
    badMode(0),

    curr_port(0),
    backPressureTimeoutNSec(20000)
{
  //add any construtor code needed for tvm pp3RGen
  pp3RxGenFiber.setRecording(TRUE);
  tbvExceptionT::setUserExceptionTypeString("pl3 CORE BACKPRESSURE");
  tbvExceptionT::setUserExceptionTypeString("pl3 ENABLE LOW TOO LONG");
}

//destructor for tvm pp3RGen
pl3RxGenTvmT::~pl3RxGenTvmT()
{
  //add any clean up code needed for tvm pp3RGen

}

//
// Method called from $tbv_tvm_connect, to create the TVM object.
//
void pl3RxGenTvmT::create ( )
{
  new pl3RxGenTvmT ( );
};
```

/***/

0.3.3   SPI4 interface

This section discusses the implementation of driver TVM for standard SPI-4 interface as recommended by Optical Internetworking Forum interface in spec titled "System Packet Interface Level 4 (SPI-4) Phase2: OC-192 System Interface for Physical and Link Layer Devices." The spec specifies the interface for interconnection of Physical layer(PHY) devices to Link Layer devices for 10Gb/s aggregate bandwidth applications. SPI-4 is an interface for packet and cell transfer between PHY device and link device, for aggregate bandwidths of OC-192 ATM and Packet over Sonet/SDH(POS), as well as 10Gb/s Ethernet applications. Following is example block diagram depicting interface signals. The transmit and receive data paths include, respectively, (TDCLK, TDAT[15:0], TDCTL) and (RDCLK, RDAT[15:0], RCTL). The transmit and receive FIFO status channels include (TSCLK, TSTAT[1:0]) and (RSCLK, RSTAT[1:0]) respectively.

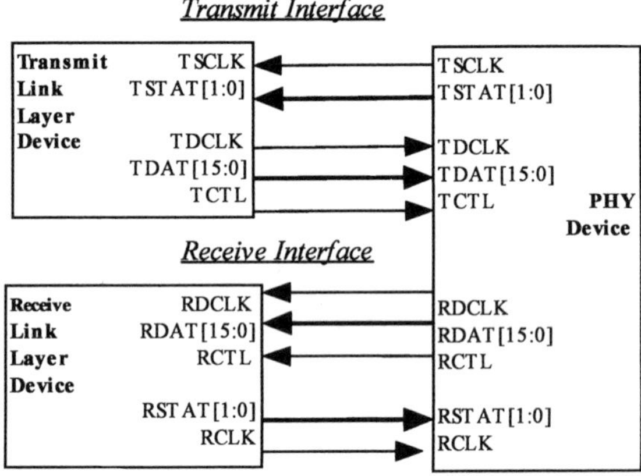

0.3.3.1   Interface Description and Operation:

This section discusses briefly about the interface operation. For more details please refer to the spec. Data is transferred in bursts that have a provisionable maximum length, with the exception of transfers that terminate with an EOP. Information associated with each transfer (port address, start/end-of-packet indication and error-control coding) is sent in 16-bit control words. Following figure shows how ATM cells and variable-length packets map onto the data stream. Complete packets or shorter bursts may be transferred as below.

Mapping of packets and ATM cells onto payload stream.

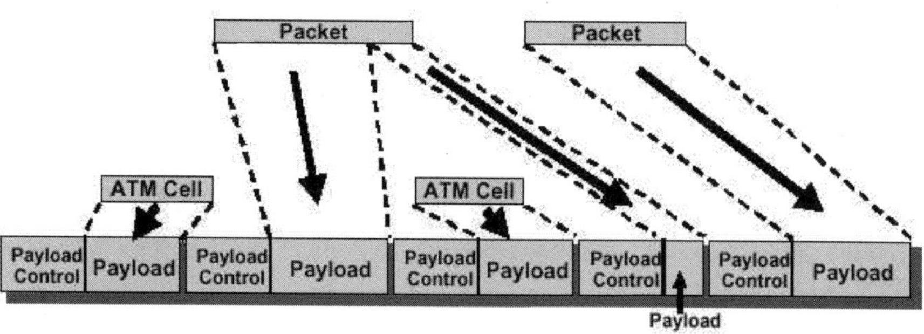

The maximum configured payload data transfer size must be a multiple of 16 bytes. Control words are inserted only between burst transfers; once a transfer has begun, data words are sent uninterrupted until end-of-packet or a multiple of 16 bytes is reached. The interval between the end of a given transfer and the next payload control word (marking the start of another transfer) consists of zero or more idle control words and training patterns.

A common control word format is used in both the transmit and receive interfaces.Various fields of control word are

- Bit 15: Control word Type (1 for payload, 0 for idle or training control word)
- Bit 14:13: End of Packet status.
  00: Not and EOP,
  01: EOP Abort
  10: EOP Normal termination(2 byte valid)
  11: EOP Normal termination (1 byte valid)
- Bit 12: Start -of -packet. Set to 1 is payload followed is start of packet.
- Bit 11:4: Port address. Set to all zeros in all idle control words and all ones in training control words
- Bit 3:0: 4 bit diagonal interleaved parity

When inserted in the data path, the control word is aligned such that its MSB is sent on the MSB of the transmit or receive data lines. A payload control word that separates two adjacent burst transfers contains status information pertaining to the previous transfer and the following transfer. Following is a list of control word type (Bits 15:12)

- 0000: Idle, not EOP, training control word
- 0001 Reserved
- 0010: Idle, Abort last packet.

For more details of control word and DIP-4 parity calculation, refer to the spec.

A timing diagram of data path signals is as shown below. This diagram is applicable to either transmit and receive interface. TCTL/RCTL is high when TDAT/RDAT contain control words. Idle periods correspond to back to back control words.

Data Path Timing Diagram

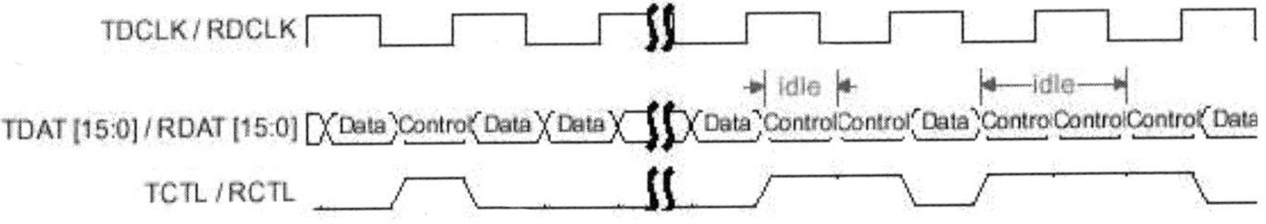

FIFO Status information is sent over TSTAT link from PHY to Link Layer Device, and over RSTA link from link Layer to the PHY device. For more information please refer to spec. Here is the timing diagram.

FIFO Status Channel Timing Diagram

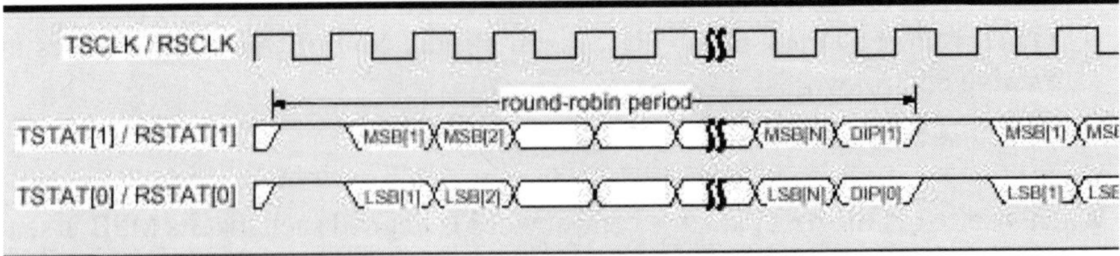

A training sequence is scheduled to be sent at least once every pre configured bounded interval on both the transmit and receive data paths. It consists of one idle control word followed by one or more repetitions od a 20-word training pattern consiting of 10(repeated) training control words and 10(repeated) training data words.

As you can see, the SPI4 interface becomes somewhat complex to implement because of lot of operations going on at one time and since payload and control is shared by same bus.

0.3.3.2 SPI4 Interface driver implementation

This section discuses implementation of driver TVM for above protocol in testBuilder using C++ based verification methodology. As described in the flow diagram below, SPI4 Tvm runs five concurrent processes:

- Process 1
  This process sets a flag (running) - that defines the period of time when the SPI Tvm can operate. This is set by a start() function.

- Process 2
  This process does status monitoring and collection (spi4TStatTask, run by runTStatTask). It has the following three main functions:

1. It runs freely and monitors the two bit status bus (while running is TRUE). |
   If the check sum DIP2 matches -> it updates the status for all ports at the same time (status comes per port for all the ports, based on a programmable number of times. Port order follows a port calendar).

2. It detects too long periods with idle status (3) and triggers training patterns (raising a flag).

3. It also detects if any port's status changes successfully (with DIP2 matching) in a favorable direction (2->1, 2->0, 1->0), so that it can trigger Data Transfer Request in the case of Non-Blocking operation, when FIFO and is not empty and there are no more spawns/runs of the TVM pending. Refer to Process 5, Non-Blocking (below).

- Process 3
  Training Pattern count (trainingCntTask) - runs freely (while running is TRUE) getting reset when a training pattern starts. If this count reaches a certain threshold, a flag is being set to trigger a training pattern.

- Process 4:
  Training Pattern (spi4PatternTask) - runs freely (while running is TRUE) check-

ing for any one of three possible flags that would require a training patterns to be sent. There are three possible reasons (flags):

1. Very beginning or Status remained idle (3) for too long. (Those 2 are combined into one flag/reason) (done by trPattern_deskewed)

2. There is no traffic and the time that has passed since last training pattern exceeds the maximum allowed. (done by trPattern_cycles)

3. There is traffic and the time passed since last training patterns has not exceeded the maximum yet, but after the end of the current transfer the maximum time will be exceeded, so a training pattern would be needed before the beginning of the current transfer. (done by trPattern_data)

The training pattern needs to differentiate between the three flags, because the control word that it has to calculate depends on the data before and/or after the training pattern.

- Process 5:

Data Task. This task sends the current transfer providing correct control words before and after the transfer. There are 2 modes of operation:

Blocking

In this case if one port is blocked for some reason, traffic is stopped for all the ports.

Generator spawns spi4TDataTask. The current status for the current port is checked. This status can be:

    0 - starving : MaxBurst1 more blocks of 16 bytes each can be sent.

    1 - hungry   : MaxBurst2 more blocks of 16 bytes each can be sent.

    2 - satisfied: no more transfers are currently allowed.

Following actions are taken

a) based on the status (0,1 or 2) select appropriate current threshold.

b) calculate the size of the current transfer in terms of 16 byte blocks.

c) calculate check sum DIP4 to be sent in the control word after the data.

d) if (data is valid) && (size < threshold) then

send the data;

    else

wait for status change and send idle data during this time.

Non-Blocking:

In this case, if one port is blocked for some reason, traffic is stopped only for that port and it continues for rest of the ports depending on port status

Generator spawns - fullStatTask.

The current transfer request is put into a FIFO. Then the whole FIFO is traversed to find a transfer request for which the status now allows the transfer to happen.

If the status allows the main task is called (the Blocking one - spi4TDataTask), but now step d) of Process 5 Blocking "if (size < threshold)" is already known to be TRUE, otherwise it would have been left in the FIFO and skipped.

If the FIFO is not empty after the last TVM spawn/run, then an additional triggering cause is necessary to execute the remaining transfer requests. This triggering cause comes from spi4TStatTask - Process 2. When status changes in a favorable way (2->1, 2->0 or 1->0) for any port another task is called (queueStatTask), which in turn again calls the main task (blocking) spi4TDataTask, having already checked the status.

Flow:

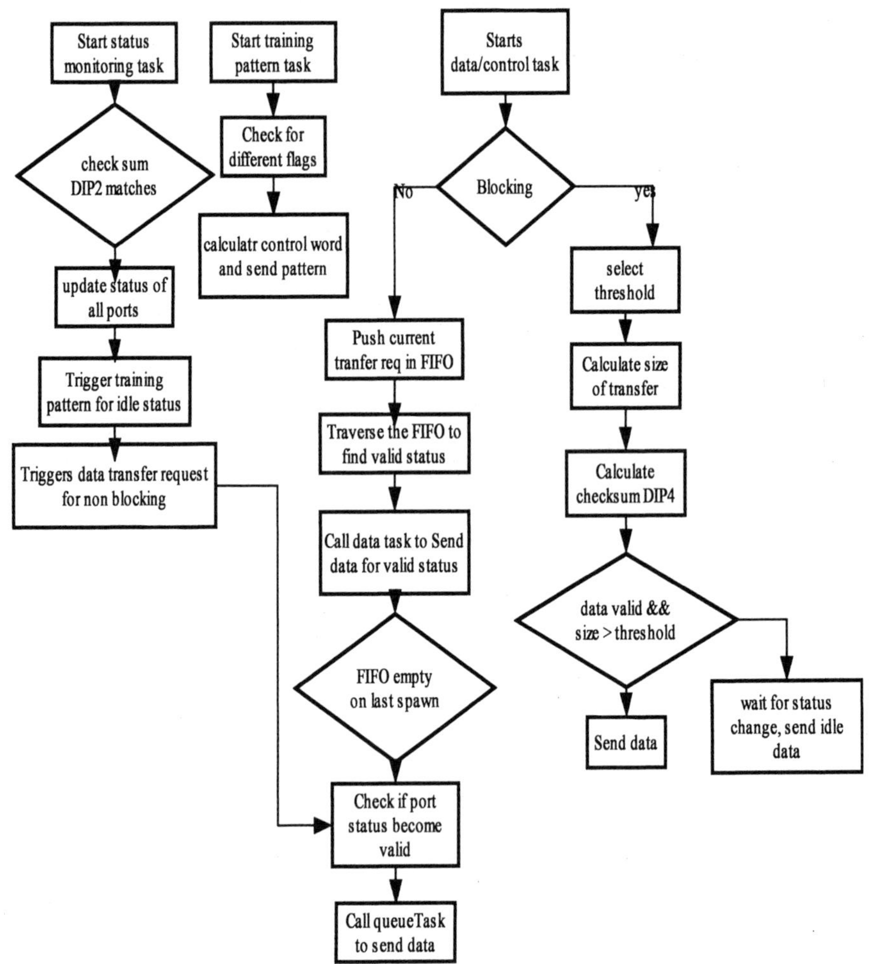

0.3.4 Code Example for SPI4 interface

spi4IfGen.h

```
#ifndef INCLUDED_SPI4IF_TVM
#define INCLUDED_SPI4IF_TVM

#include "TestBuilder.h"
#include "tbvCell.h"
//#include <stl.h>
//#include <queue>
```

```cpp
class spi4IfTvmT;


// *************************************
// a task spi4TData
// *************************************
class spi4TDataTaskT : public tbvTaskTypeSafeT<tbvCellT>
{
  public:
    //parent tvm, make it easy to access in task body
    spi4IfTvmT& spi4IfTvm;

    //constructor, destructor
    spi4TDataTaskT(spi4IfTvmT& );
    virtual ~spi4TDataTaskT();

    //body function for task functionality
    virtual void body(tbvCellT *TrGenArgsP );

    void trPattern_data();
};


// *************************************
// a task fullStat
// *************************************
class fullStatTaskT : public tbvTaskTypeSafeT<tbvCellT>
{
  private:
    spi4IfTvmT& spi4IfTvm;
    spi4TDataTaskT& mainTask;
    //vector<queue<tbvCellT *> > portQueue;

  public:
    fullStatTaskT(spi4IfTvmT& spi4IfTvm, spi4TDataTaskT & mainTask);
    virtual ~fullStatTaskT();

    //body function for task functionality
    virtual void body(tbvCellT *TrGenArgsP );
};


// *************************************
// a task queueStat
// *************************************
class queueStatTaskT : public tbvTaskT
{
  private:
    spi4IfTvmT& spi4IfTvm;
    spi4TDataTaskT& mainTask;

  public:
    queueStatTaskT(spi4IfTvmT& spi4IfTvm, spi4TDataTaskT & mainTask);
    virtual ~queueStatTaskT();

    //body function for task functionality
    virtual void body(tbvSmartDataT * );
};


// *************************************
```

```cpp
// a task spi4Pattern
// ************************************
class spi4PatternTaskT : public tbvTaskTypeSafeT<tbvCellT>
{
  public:
    //parent tvm, make it easy to access in task body
    spi4IfTvmT& spi4IfTvm;

    //constructor, destructor
    spi4PatternTaskT(spi4IfTvmT& );
    virtual ~spi4PatternTaskT();

    //body function for task functionality
    virtual void body(tbvCellT *TrGenArgsP );

    void trPattern_deskewed();
    void trPattern_cycles();

};


// ****************************************
// a task spi4TStat
// ****************************************
class spi4TStatTaskT : public tbvTaskTypeSafeT<tbvCellT>
{
  public:
    //parent tvm, make it easy to access in task body
    spi4IfTvmT& spi4IfTvm;

    //constructor, destructor
    spi4TStatTaskT(spi4IfTvmT& );
    virtual ~spi4TStatTaskT();

    //body function for task functionality
    virtual void body(tbvCellT *TrGenArgsP );
};


// ****************************************
// a task runTStat
// ****************************************
class runTStatTaskT : public tbvTaskT
{
  public:
    //parent tvm, make it easy to access in task body
    spi4IfTvmT& spi4IfTvm;

    //constructor, destructor
    runTStatTaskT(spi4IfTvmT& );
    virtual ~runTStatTaskT();

    //body function for task functionality
    virtual void body(tbvSmartDataT * );
};


// ****************************************
// a task trainingCnt
// ****************************************
class trainingCntTaskT : public tbvTaskT
{
```

```cpp
public:
  //parent tvm, make it easy to access in task body
  spi4IfTvmT& spi4IfTvm;

  //constructor, destructor
  trainingCntTaskT(spi4IfTvmT& );
  virtual ~trainingCntTaskT();

  //body function for task functionality
  virtual void body(tbvSmartDataT * );
};


// ***************************************
// a TVM spi4IfTvm
// ***************************************
class spi4IfTvmT : public tbvTvmT
{
public:
  unsigned int status[256];
  unsigned int port_cal[256];
  unsigned int block16_cnt[256];
  unsigned int fl[256];
  unsigned int dip2;
  unsigned int curr_eop;
  unsigned int odd_byte;
  unsigned int dip4;
  unsigned int block_16;
  unsigned int thresh[256];

  //
  // These signals are connected to Verilog signals
  // through the test builder registry
  //
  tbvSignalHdlT tdclk;
  tbvSignalHdlT tsclk;
  tbvSignalHdlT reset;
  tbvSignalHdlT tdat;
  tbvSignalHdlT tctl;
  tbvSignalHdlT tstat;

  // The Mutex
  tbvMutexT spi4TDataMutex;
  tbvMutexT fullStatMutex;
  tbvMutexT queueStatMutex;
  tbvMutexT spi4TStatMutex;
  tbvMutexT portQueueMutex;

  tbvFiberT spi4IfFiber;
  tbvFiberT fullStatFiber;
  tbvFiberT queueStatFiber;

  //
  // basic tasks
  //
  spi4TDataTaskT spi4TData;
  fullStatTaskT fullStat;
  queueStatTaskT queueStat;
  spi4PatternTaskT spi4Pattern;
  spi4TStatTaskT spi4TStat;
  runTStatTaskT runTStat;
  trainingCntTaskT trainingCnt;
```

```cpp
    bool running;
    void start(){running = true;}
    void stop();

    unsigned int cal_len;
    unsigned int cal_m;

    void setCal_len(unsigned int k){cal_len = k;}
    void setCal_m(unsigned int k){cal_m = k;}
    vector <uint16> old_data;
    vector <uint16> MaxBurst1;
    vector <uint16> MaxBurst2;
    unsigned int invalid;
    unsigned int last_data_valid;
    unsigned int last_data_bad;
    unsigned int deskewed;

    unsigned int flag_data;
    unsigned int flag_deskewed;
    unsigned int flag_cycles;
    unsigned int size4pattern;
    unsigned int training_count;
    unsigned int data_max_t;
    unsigned int training_m;
    void setData_Max_T(unsigned int k){data_max_t = k;}
    void setTraining_M(unsigned int k){training_m = k;}

    unsigned int badParity;
    void setBadParity(unsigned int k){badParity = k;}

    unsigned int fullPort;
    void setFullPort(unsigned int k){fullPort = k;}

    vector<tbvCellT *> portQueue;

    unsigned int dip4_calc(unsigned int, unsigned int, vector <uint16> &);


    // Event Expressions
    tbvEventExprT anyedge_clk;
    tbvEventExprT posedge_tsclk;
    tbvEventExprT not_reset;

    //
    // This function is called from $tbv_tvm_connect, to instantiate this TVM.
    //
    static void create();

    //
    // Constructor / Destructor
    //
    spi4IfTvmT();
    virtual ~spi4IfTvmT();
};


#endif


/********************************************************/
```
spi4IfGen.cc

```cpp
#include "TestBuilder.h"
#include "spi4IfGen.h"


// **************************************
// a task spi4TData
// **************************************
spi4TDataTaskT::spi4TDataTaskT(spi4IfTvmT& spi4IfTvm)
  : tbvTaskTypeSafeT<tbvCellT>(&spi4IfTvm, "spi4TDataTask"),
    spi4IfTvm(spi4IfTvm)

{
  // Disable the mutex built into the fiber and turn off automatic
  // transaction recording
  setSynchronization(FALSE, FALSE);
}


spi4TDataTaskT::~spi4TDataTaskT() {}

  //
  // TVM spi4IfTvm : body method for task spi4TData
  //

void spi4TDataTaskT::body(tbvCellT *TrGenArgsP)
{

  //
  // Do one transaction at a time
  //

  spi4IfTvm.spi4TDataMutex.lock();

  //
  // wait for Reset to end (if it is asserted)
  //

  if (tbv4StateNotEqual(spi4IfTvm.reset, 1))
    tbvWait( spi4IfTvm.not_reset && spi4IfTvm.anyedge_clk , tbvEventExprT::RWSYNC );

  //For random bad parity generation
  tbvSmartUnsignedT badPrtyRand;
  badPrtyRand.keepOnly(1,100);
  badPrtyRand.randomize();


  //For Pattern task - calculate the number of clock cycles the current burst will take
  if ((TrGenArgsP->getSize())%2 == 0)
    spi4IfTvm.size4pattern = (TrGenArgsP->getSize())/2;
  else
    spi4IfTvm.size4pattern = (TrGenArgsP->getSize())/2 + 1;

  spi4IfTvm.size4pattern = spi4IfTvm.size4pattern + 1;  // one clk cycle for cntr word

  // If the number of clocks since last training pattern + the clocks the current burst will need exceeds
  the programmed max threshold ->run a training pattern
  if ((spi4IfTvm.size4pattern + spi4IfTvm.training_count) > spi4IfTvm.data_max_t)
    {
      spi4IfTvm.training_count = 0;
      trPattern_data();
    }
```

```cpp
// Start transaction recording
spi4IfTvm.spi4IfFiber.beginTransactionH("spi4IfFiber");

vector<uint16> data;

// Take previous data - needed for calculation of DIP4 later on
data = spi4IfTvm.old_data;
// Update the previous data with the current one
spi4IfTvm.old_data = TrGenArgsP->getDataBy16();

// Get the current burst size in terms of 16-byte blocks
if ((TrGenArgsP->getSize())%16 == 0)
   spi4IfTvm.block_16 = (TrGenArgsP->getSize())/16;
else
   spi4IfTvm.block_16 = (TrGenArgsP->getSize())/16 + 1;

// Based on the status of the current port select the appropriate threshold :
// 0 - MaxBurst1 starving; 1 - MaxBusrt2 hungry; 2 - satisfied;
if (spi4IfTvm.status[TrGenArgsP->getInputPort()] == 0)
   spi4IfTvm.thresh[TrGenArgsP->getInputPort()] = spi4IfTvm.MaxBurst1[TrGenArgsP->getInput-
Port()];
else if (spi4IfTvm.status[TrGenArgsP->getInputPort()] == 1)
   spi4IfTvm.thresh[TrGenArgsP->getInputPort()] = spi4IfTvm.MaxBurst2[TrGenArgsP->getInput-
Port()];
else {
   tbvExceptionT::setUserExceptionTypeString("SPI4 CORE BACKPRESSURE");
   tbvExceptionT::reportUserException("SPI4 CORE BACKPRESSURE",
      tbvExceptionT::WARNING,
      "Backpressure on port %d @ %d ns",(TrGenArgsP->getInputPort()).getUnsignedValue(), (int)
tbvGetInt64Time(TBV_NS));
   spi4IfTvm.thresh[TrGenArgsP->getInputPort()] = 0;
}

// Odd Parity Calculation

   unsigned int data_cntr;
   unsigned int data_cntr_while;
   unsigned int dip16_while;

   data_cntr = 0;
   data_cntr_while = 0;

// control word for both valid and inalid data
   data_cntr = ((((spi4IfTvm.block16_cnt[TrGenArgsP->getInputPort()] + spi4IfTvm.block_16) >
spi4IfTvm.thresh[TrGenArgsP->getInputPort()]) ?  0 : (TrGenArgsP->getValid())) << 15) |
(spi4IfTvm.curr_eop << 14) | (spi4IfTvm.odd_byte << 13) | (((((TrGenArgsP->getValid() == 0) ||
(spi4IfTvm.block16_cnt[TrGenArgsP->getInputPort()] + spi4IfTvm.block_16) >
spi4IfTvm.thresh[TrGenArgsP->getInputPort()])) ? 0 : (TrGenArgsP->getSOP())) << 12) | (((((TrGe-
nArgsP->getValid() == 0) || (spi4IfTvm.block16_cnt[TrGenArgsP->getInputPort()] +
spi4IfTvm.block_16) > spi4IfTvm.thresh[TrGenArgsP->getInputPort()]))? 0 : (TrGenArgsP->get-
InputPort())) << 4) | 0xf;

   data_cntr_while = (TrGenArgsP->getValid() << 15) | (0 << 14) | (0 << 13) | (((TrGenArgsP-
>getValid() == 0) ? 0 : (TrGenArgsP->getSOP())) << 12) | (((TrGenArgsP->getValid() == 0) ? 0 :
(TrGenArgsP->getInputPort())) << 4) | 0xf;

   dip16_while = data_cntr_while;

   spi4IfTvm.dip4 = spi4IfTvm.dip4_calc(0, data_cntr, data);

   unsigned int  dip4_while =
```

```
        (((((dip16_while & (0x8000 >> 0)) ? 1 : 0) ^ ((dip16_while & (0x8000 >> 4)) ? 1 : 0) ^ ((dip16_while &
      (0x8000 >> 8)) ? 1 : 0) ^ ((dip16_while & (0x8000 >> 12)) ? 1 : 0)) << 3) |
        (((((dip16_while & (0x8000 >> 1)) ? 1 : 0) ^ ((dip16_while & (0x8000 >> 5)) ? 1 : 0) ^ ((dip16_while &
      (0x8000 >> 9)) ? 1 : 0) ^ ((dip16_while & (0x8000 >> 13)) ? 1 : 0)) << 2) |
        (((((dip16_while & (0x8000 >> 2)) ? 1 : 0) ^ ((dip16_while & (0x8000 >> 6)) ? 1 : 0) ^ ((dip16_while
      & (0x8000 >> 10)) ? 1 : 0) ^ ((dip16_while & (0x8000 >> 14)) ? 1 : 0)) << 1) |
        ((((dip16_while & (0x8000 >> 3)) ? 1 : 0) ^ ((dip16_while & (0x8000 >> 7)) ? 1 : 0) ^ ((dip16_while
      & (0x8000 >> 11)) ? 1 : 0) ^ ((dip16_while & (0x8000 >> 15)) ? 1 : 0));

      unsigned int was_in_while = 0;

    // While we have valid, but the current transfer exceeds the threshold we have to send invalid data
    with a control word representing
    // valid or invalid data before this point.

      while (((TrGenArgsP->getValid() == 1) && ((spi4IfTvm.block16_cnt[TrGenArgsP->getInputPort()] +
    spi4IfTvm.block_16) > spi4IfTvm.thresh[TrGenArgsP->getInputPort()])))
        {
          if (spi4IfTvm.block_16 > spi4IfTvm.MaxBurst1[TrGenArgsP->getInputPort()]) {
            tbvExceptionT::setUserExceptionTypeString("SPI4 MAX BURST VIOLATION");
            tbvExceptionT::reportUserException("SPI4 MAX BURST VIOLATION",
    tbvExceptionT::ERROR,
    "For port %d, Burst size = %d > MaxBurst1 = %d",
    (unsigned) TrGenArgsP->getInputPort(), spi4IfTvm.block_16,
    spi4IfTvm.MaxBurst1[TrGenArgsP->getInputPort()]);
            tbvExit();
          }
    // define/send control word before the invalid data
        if (spi4IfTvm.last_data_valid)
          {
            spi4IfTvm.tctl = 1;
            spi4IfTvm.tdat(15,15) = 0;
    if (spi4IfTvm.curr_eop)
     if (spi4IfTvm.last_data_bad) {
       spi4IfTvm.tdat(14,13) = 1; // EOP Abort
       if (spi4IfTvm.odd_byte != 1)
         spi4IfTvm.dip4 ^= 0x6;
       else
         spi4IfTvm.dip4 ^= 0x4;
     }
     else {
       spi4IfTvm.tdat(14,14) = 1; // EOP Normal termination
       spi4IfTvm.tdat(13,13) = spi4IfTvm.odd_byte; // 0 = 2 bytes valid, 1 = 1 byte valid
     }
    else
     spi4IfTvm.tdat(14,13) = 0; // Not EOP

          spi4IfTvm.curr_eop = 0;
          spi4IfTvm.odd_byte = 0;
          spi4IfTvm.tdat(12,12) = 0;
          spi4IfTvm.tdat(11,4) = 0;
          if (badPrtyRand < spi4IfTvm.badParity) {
    tbvOut << spi4IfTvm.getInstanceNameP() << "generating bad parity cell\n";
            spi4IfTvm.tdat(3,0) = spi4IfTvm.dip4 ^ 1;
    }
          else
            spi4IfTvm.tdat(3,0) = spi4IfTvm.dip4;

          tbvWaitCycle(spi4IfTvm.tdclk, tbvThreadT::ANYEDGE);

        }
```

```
// Sending invalid data
   spi4IfTvm.tdat = 0x000f;
   tbvWaitCycle(spi4IfTvm.tdclk, tbvThreadT::ANYEDGE);
   spi4IfTvm.last_data_valid = 0;
   was_in_while = 1;

// Keep updating the status while waiting and sending invalid data
   if (spi4IfTvm.status[TrGenArgsP->getInputPort()] == 0)
     spi4IfTvm.thresh[TrGenArgsP->getInputPort()] = spi4IfTvm.MaxBurst1[TrGenArgsP->getInput-
Port()];
   else
     if (spi4IfTvm.status[TrGenArgsP->getInputPort()] == 1)
       spi4IfTvm.thresh[TrGenArgsP->getInputPort()] = spi4IfTvm.MaxBurst2[TrGenArgsP->get-
InputPort()];
     else
       spi4IfTvm.thresh[TrGenArgsP->getInputPort()] = 0;
  }

// Here we have enough credit to send the current transfer
// Sending control word and valid data
  if (TrGenArgsP->getValid() == 1)
  {
    spi4IfTvm.tctl = 1;

    // Send the controls
    spi4IfTvm.tdat(15,15) = 1;
    if (spi4IfTvm.curr_eop)
     if (spi4IfTvm.last_data_bad) {
spi4IfTvm.tdat(14,13) = 1; // EOP Abort
if (spi4IfTvm.odd_byte != 1)
  spi4IfTvm.dip4 ^= 0x6;
else
  spi4IfTvm.dip4 ^= 0x4;
     }
     else {
spi4IfTvm.tdat(14,14) = 1; // EOP Normal termination
spi4IfTvm.tdat(13,13) = spi4IfTvm.odd_byte; // 0 = 2 bytes valid, 1 = 1 byte valid
     }
    else
      spi4IfTvm.tdat(14,13) = 0; // Not EOP

    spi4IfTvm.curr_eop = TrGenArgsP->getEOP();
    spi4IfTvm.odd_byte = (TrGenArgsP->getSize())%2;
    spi4IfTvm.tdat(12,12) = TrGenArgsP->getSOP();
    spi4IfTvm.tdat(11,4) = TrGenArgsP->getInputPort();
    spi4IfTvm.fl[TrGenArgsP->getInputPort()] = 0;

    if(spi4IfTvm.invalid) {
     if (badPrtyRand < spi4IfTvm.badParity) {
tbvOut << spi4IfTvm.getInstanceNameP() << "generating bad parity cell\n";
       spi4IfTvm.tdat(3,0) = dip4_while ^ 1; //0xF;
     }
     else
       spi4IfTvm.tdat(3,0) = dip4_while;
    }
    else{
     if (badPrtyRand < spi4IfTvm.badParity) {
tbvOut << spi4IfTvm.getInstanceNameP() << "generating bad parity cell\n";
       spi4IfTvm.tdat(3,0) = (was_in_while) ? (dip4_while ^ 1) : (spi4IfTvm.dip4 ^ 1);
     }
     else
       spi4IfTvm.tdat(3,0) = (was_in_while) ? (dip4_while) : (spi4IfTvm.dip4);
```

```cpp
      }

    tbvWaitCycle(spi4IfTvm.tdclk, tbvThreadT::ANYEDGE);
    spi4IfTvm.tctl = 0;

    data = spi4IfTvm.old_data;

    int beginTime;

    // Send the data
    for (unsigned int i = 0; i < data.size(); i++)
      {
        if ((i == data.size()-1) && (spi4IfTvm.odd_byte == 1))
          spi4IfTvm.tdat = data[i]&0xff00;
        else
          spi4IfTvm.tdat = data[i];

        tbvWaitCycle(spi4IfTvm.tdclk, tbvThreadT::ANYEDGE);

        beginTime = tbvGetInt64Time(TBV_NS);
      }

    spi4IfTvm.block16_cnt[TrGenArgsP->getInputPort()] = (spi4IfTvm.block16_cnt[TrGenArgsP-
>getInputPort()] + spi4IfTvm.block_16);

    spi4IfTvm.tctl = 1;
    spi4IfTvm.invalid = 0;
    spi4IfTvm.last_data_valid = 1;

  }

  else

// Sending invalid data
  {
    spi4IfTvm.tctl = 1;
    spi4IfTvm.tdat(15,15) = 0;
    if (spi4IfTvm.curr_eop)
      if (spi4IfTvm.last_data_bad) {
spi4IfTvm.tdat(14,13) = 1; // EOP Abort
if (spi4IfTvm.odd_byte != 1)
  spi4IfTvm.dip4 ^= 0x6;
else
  spi4IfTvm.dip4 ^= 0x4;
    }
      else {
spi4IfTvm.tdat(14,14) = 1; // EOP Normal termination
spi4IfTvm.tdat(13,13) = spi4IfTvm.odd_byte; // 0 = 2 bytes valid, 1 = 1 byte valid
    }
    else
      spi4IfTvm.tdat(14,13) = 0; // Not EOP

    spi4IfTvm.curr_eop = 0;
    spi4IfTvm.odd_byte = 0;
    spi4IfTvm.tdat(12,12) = 0;
    spi4IfTvm.tdat(11,4) = 0;
 // spi4IfTvm.tdat(3,0) = 0;

    if (spi4IfTvm.last_data_valid == 0)
      spi4IfTvm.tdat(3,0) = 0xf;
    else
      {
```

```cpp
if (badPrtyRand < spi4IfTvm.badParity) {
  tbvOut << spi4IfTvm.getInstanceNameP() << "generating bad parity cell\n";
      spi4IfTvm.tdat(3,0) = spi4IfTvm.dip4 ^ 1;
}
    else
      spi4IfTvm.tdat(3,0) = spi4IfTvm.dip4;
    }

    tbvWaitCycle(spi4IfTvm.tdclk, tbvThreadT::ANYEDGE);


    data = TrGenArgsP->getDataBy16();

    // Send the data
    for (unsigned int i = 0; i < data.size()+1; i++)
      {
        spi4IfTvm.tdat = 0x000f;  //data[i];

        tbvWaitCycle(spi4IfTvm.tdclk, tbvThreadT::ANYEDGE);
      }

    vector<uint16> invalid(data.size(),0x000f);
    spi4IfTvm.old_data = invalid;
    spi4IfTvm.invalid = 1;
    spi4IfTvm.last_data_valid = 0;

  }


  spi4IfTvm.last_data_bad = TrGenArgsP->getBad();
  // Record the values in the argument block onto the fiber
  spi4IfTvm.spi4IfFiber.recordAttribute(*TrGenArgsP, "spi4IfFiber");

  // Finish Transaction Recording
  spi4IfTvm.spi4IfFiber.endTransaction();


  spi4IfTvm.spi4TDataMutex.unlock();

}

// ************************************
// trPattern_data function
// ************************************
void spi4TDataTaskT::trPattern_data()
{
  //spi4IfTvm.spi4PatternMutex.lock();
  //spi4IfTvm.spi4TDataMutex.lock();

  //
  // wait for Reset to end (if it is asserted)
  //
  if (tbv4StateNotEqual(spi4IfTvm.reset, 1))
    tbvWait( spi4IfTvm.not_reset && spi4IfTvm.anyedge_clk , tbvEventExprT::RWSYNC );


  // Start transaction recording
  spi4IfTvm.spi4IfFiber.beginTransactionH("training pattern");

  spi4IfTvm.flag_data = 1;

// vector<uint16> data = TrGenArgsP->getDataBy16();
```

```cpp
vector<uint16> data;

data = spi4IfTvm.old_data;
//spi4IfTvm.old_data = TrGenArgsP->getDataBy16();


// Odd Parity Calculation - calculated similarly to Odd Parity Calculation in spi4TDataTask

unsigned int data_cntr;

data_cntr = 0;

data_cntr = (0 << 15) | (spi4IfTvm.curr_eop << 14) | (spi4IfTvm.odd_byte << 13) | (0 << 12) | (0 <<
4) | 0xf;

spi4IfTvm.dip4 = spi4IfTvm.dip4_calc(spi4IfTvm.invalid, data_cntr, data);


// Send the control word before the training pattern

spi4IfTvm.tctl = 1;
spi4IfTvm.tdat(15,15) = 0;
spi4IfTvm.tdat(14,14) = (spi4IfTvm.last_data_valid) ? spi4IfTvm.curr_eop : 0;
spi4IfTvm.tdat(13,13) = (spi4IfTvm.last_data_valid) ? spi4IfTvm.odd_byte : 0;
spi4IfTvm.curr_eop = 0;
spi4IfTvm.odd_byte = 0;
spi4IfTvm.tdat(12,12) = 0;
spi4IfTvm.tdat(11,4) = 0;
spi4IfTvm.tdat(3,0) = spi4IfTvm.dip4;
//spi4IfTvm.tdat(3,0) = (spi4IfTvm.last_data_valid) ? spi4IfTvm.dip4 : dip4_while;

//tbvOut << " DIP4_PATTERN at : " << spi4IfTvm.tdat(3,0) << tbvGetInt64Time(TBV_NS) << " ns"
<< endl;

tbvWaitCycle(spi4IfTvm.tdclk, tbvThreadT::ANYEDGE);

// Send the pattern
for (unsigned int m2 = 0; m2 < spi4IfTvm.training_m; m2++)
for (unsigned int i = 0; i < 10; i++)
  {
    spi4IfTvm.tctl = 1;
    spi4IfTvm.tdat = 0x0fff;

    for (unsigned int k = 0; k < 10; k++)
      tbvWaitCycle(spi4IfTvm.tdclk, tbvThreadT::ANYEDGE);

    spi4IfTvm.tctl = 0;
    spi4IfTvm.tdat = 0xf000;

    for (unsigned int j = 0; j < 10; j++)
      tbvWaitCycle(spi4IfTvm.tdclk, tbvThreadT::ANYEDGE);
  }

spi4IfTvm.training_count = 0;
spi4IfTvm.tctl = 1;
spi4IfTvm.tdat = 0x000f;
spi4IfTvm.last_data_valid = 0;
spi4IfTvm.invalid = 1;

spi4IfTvm.flag_data = 0;

// Record the values in the argument block onto the fiber
```

```cpp
    tbvCellT invalid_pattern(1);
    spi4IfTvm.spi4IfFiber.recordAttribute(invalid_pattern, "training pattern");

    // Finish Transaction Recording
    spi4IfTvm.spi4IfFiber.endTransaction();


    //spi4IfTvm.spi4PatternMutex.unlock();
    //spi4IfTvm.spi4TDataMutex.unlock();

}


// ***************************************
// a task fullStat
// ***************************************
fullStatTaskT::fullStatTaskT(spi4IfTvmT& spi4IfTvm, spi4TDataTaskT& mainTask)
  : tbvTaskTypeSafeT<tbvCellT>(&spi4IfTvm, "fullStatTask"),
    spi4IfTvm(spi4IfTvm),
    mainTask(mainTask)

{
  // Disable the mutex built into the fiber and turn off automatic
  // transaction recording
  //setSynchronization(FALSE, FALSE);
}


fullStatTaskT::~fullStatTaskT() {}

  //
  // TVM spi4IfTvm : body method for task fullStat
  //

void fullStatTaskT::body(tbvCellT *TrGenArgsP)
{

  //
  // Do one transaction at a time
  //

  spi4IfTvm.fullStatMutex.lock();
  spi4IfTvm.portQueueMutex.lock();

  //
  // wait for Reset to end (if it is asserted)
  //

  if (tbv4StateNotEqual(spi4IfTvm.reset, 1))
    tbvWait( spi4IfTvm.not_reset && spi4IfTvm.anyedge_clk , tbvEventExprT::RWSYNC );

  // Push the current transfer request into the FIFO
  spi4IfTvm.portQueue.push_back(TrGenArgsP);

  tbvCellT *pipedArg = NULL;

  unsigned int i = 0;
  unsigned int main_task_run = 0;
  unsigned int flag_stat[64];

  for (unsigned int i = 0; i < 64; i++)
    flag_stat[i] = 0;
```

```cpp
// Go through all FIFO entries
while (i < spi4IfTvm.portQueue.size())
 if (spi4IfTvm.portQueue.size() > 0)
  {
    pipedArg = spi4IfTvm.portQueue[i];

// Calculate the number of 16-byte chunks in the current transfer
    if ((pipedArg->getSize())%16 == 0)
      spi4IfTvm.block_16 = (pipedArg->getSize())/16;
    else
      spi4IfTvm.block_16 = (pipedArg->getSize())/16 + 1;

// Select the correct threshold based on the status for the current port
    if (spi4IfTvm.status[pipedArg->getInputPort()] == 0)
      spi4IfTvm.thresh[pipedArg->getInputPort()] = spi4IfTvm.MaxBurst1[pipedArg->getInputPort()];
    else
     if (spi4IfTvm.status[pipedArg->getInputPort()] == 1)
       spi4IfTvm.thresh[pipedArg->getInputPort()] = spi4IfTvm.MaxBurst2[pipedArg->getInputPort()];
     else
       spi4IfTvm.thresh[pipedArg->getInputPort()] = 0;

    flag_stat[pipedArg->getInputPort().getUnsignedValue()] = flag_stat[pipedArg->getInput-
Port().getUnsignedValue()] | (((spi4IfTvm.block_16 + spi4IfTvm.block16_cnt[pipedArg->getInput-
Port()]) < spi4IfTvm.thresh[pipedArg->getInputPort()]) ? 0 : 1);

// Send the current transfer request if the threshold condition is met or if the data is invalid (no need
to check threshold)
    if ((((spi4IfTvm.block_16 + spi4IfTvm.block16_cnt[pipedArg->getInputPort()]) <
spi4IfTvm.thresh[pipedArg->getInputPort()]) && (flag_stat[pipedArg->getInputPort().getUnsigned-
Value()] == 0)) || (pipedArg->getValid() == 0))
     {
       if (spi4IfTvm.portQueue.size() > 0)
        {
          mainTask.run(pipedArg);
          main_task_run = 1;
        }
       //spi4IfTvm.portQueue.erase[i];
       if (spi4IfTvm.portQueue.size() > 0)
         {
// Shift up all entirs in the FIFO and erase the last one (erase did not work)
          for (unsigned int j = i; j < spi4IfTvm.portQueue.size() - 1; j++)
            {
              spi4IfTvm.portQueue[j] = spi4IfTvm.portQueue[j+1];
            }
          spi4IfTvm.portQueue.pop_back();
          i = i - 1;
         }
      }
    i = i + 1;
  }
 // send invalid cell if nothing can be popped from the FIFO
tbvCellT invalid_cell;
vector<uint8> pkt(64);
tbvControlDataT control;
control.fid.input_port = 0;
control.valid = 0;
control.sop = 0;
control.eop = 0;

for (int j = 0; j < 64; j++)
   pkt[j] = j;
```

```cpp
  invalid_cell.setPayload(pkt);
  invalid_cell.setControlData(control);

  if (main_task_run == 0)
   {
     mainTask.run(&invalid_cell);
     //tbvOut << "INVALID CELL" << endl;
   }

     spi4IfTvm.portQueueMutex.unlock();
     spi4IfTvm.fullStatMutex.unlock();


 }


// ********************************************************************
// a task queueStat
// ********************************************************************
queueStatTaskT::queueStatTaskT (spi4IfTvmT& spi4IfTvm, spi4TDataTaskT& mainTask)
  : tbvTaskT ( &spi4IfTvm, "queueStatTask" ),
    spi4IfTvm ( spi4IfTvm ),
    mainTask(mainTask)

{
  // Disable the mutex built into the fiber and turn off automatic
  // transaction recording
  setSynchronization(FALSE, FALSE);
}

queueStatTaskT::~queueStatTaskT () {}

void queueStatTaskT::body ( tbvSmartDataT *TrGenArgsP )
{

  spi4IfTvm.queueStatMutex.lock();
  spi4IfTvm.portQueueMutex.lock();


  if (tbv4StateNotEqual(spi4IfTvm.reset, 1))
     tbvWait( spi4IfTvm.not_reset && spi4IfTvm.anyedge_clk , tbvEventExprT::RWSYNC );

  // send invalid cell if nothing can be popped from the FIFO
  tbvCellT last_invalid_cell;
  vector<uint8> pkt(64);
  tbvControlDataT control;
  control.fid.input_port = 0;
  control.valid = 0;
  control.sop = 0;
  control.eop = 0;

  for (int j = 0; j < 64; j++)
      pkt[j] = j;
  last_invalid_cell.setPayload(pkt);
  last_invalid_cell.setControlData(control);


  tbvCellT *pipedArg = NULL;

  unsigned int i = 0;
  unsigned int sent_successfully = 0;
  unsigned int stat_flag[64];
```

```cpp
for (unsigned int i = 0; i < 64; i++)
   stat_flag[i] = 0;

// Go through all FIFO entries
while (i < spi4IfTvm.portQueue.size())
 if (spi4IfTvm.portQueue.size() > 0)
  {
    pipedArg = spi4IfTvm.portQueue[i];

// Calculate the number of 16-byte chunks in the current transfer
    if ((pipedArg->getSize())%16 == 0)
      spi4IfTvm.block_16 = (pipedArg->getSize())/16;
    else
      spi4IfTvm.block_16 = (pipedArg->getSize())/16 + 1;


// Select the correct threshold based on the status for the current port
    if (spi4IfTvm.status[pipedArg->getInputPort()] == 0)
      spi4IfTvm.thresh[pipedArg->getInputPort()] = spi4IfTvm.MaxBurst1[pipedArg->getInputPort()];
    else
      if (spi4IfTvm.status[pipedArg->getInputPort()] == 1)
        spi4IfTvm.thresh[pipedArg->getInputPort()] = spi4IfTvm.MaxBurst2[pipedArg->getInputPort()];
      else
        spi4IfTvm.thresh[pipedArg->getInputPort()] = 0;


    stat_flag[pipedArg->getInputPort().getUnsignedValue()] = stat_flag[pipedArg->getInput-
Port().getUnsignedValue()] | (((spi4IfTvm.block_16 + spi4IfTvm.block16_cnt[pipedArg->getInput-
Port()]) < spi4IfTvm.thresh[pipedArg->getInputPort()]) ? 0 : 1);

// Send the current transfer request if the threshold condition is met or if the data is invalid (no need
to check threshold)
    if (((((spi4IfTvm.block_16 + spi4IfTvm.block16_cnt[pipedArg->getInputPort()]) <
spi4IfTvm.thresh[pipedArg->getInputPort()]) && (stat_flag[pipedArg->getInputPort().getUnsigned-
Value()] == 0)) || (pipedArg->getValid() == 0))
      {
        if (spi4IfTvm.portQueue.size() > 0)
         {
           mainTask.run(pipedArg);
           sent_successfully = 1;
         }
        if (spi4IfTvm.portQueue.size() > 0)
         {
// Shift up all entirs in the FIFO and erase the last one (erase did not work)
           for (unsigned int j = i; j < spi4IfTvm.portQueue.size() - 1; j++)
            {
              spi4IfTvm.portQueue[j] = spi4IfTvm.portQueue[j+1];
            }
           spi4IfTvm.portQueue.pop_back();
           i = i - 1;
         }
      }
    i = i + 1;
  }

//Send Last Invalid cell to match the last parity
  if(sent_successfully == 1)
    {
      mainTask.run(&last_invalid_cell);
      tbvOut << "LAST_INVALID CELL" << endl;
    }
```

```cpp
    spi4IfTvm.portQueueMutex.unlock();
    spi4IfTvm.queueStatMutex.unlock();

}


// ***************************************
// a task spi4Pattern
// ***************************************
spi4PatternTaskT::spi4PatternTaskT(spi4IfTvmT& spi4IfTvm)
 : tbvTaskTypeSafeT<tbvCellT>(&spi4IfTvm, "spi4PatternTask"),
   spi4IfTvm(spi4IfTvm)

{
  // Disable the mutex built into the fiber and turn off automatic
  // transaction recording
  setSynchronization(FALSE, FALSE);
 }


spi4PatternTaskT::~spi4PatternTaskT() {}


 //
 // TVM spi4IfTvm : body method for task spi4Pattern
 //

void spi4PatternTaskT::body(tbvCellT *TrGenArgsP)
{

while (spi4IfTvm.running)
{
// Send training pattern if just starting simulation or if Status = 3 for
// too long or training ocunter has reached its max value
 if ((spi4IfTvm.deskewed) || (spi4IfTvm.training_count == spi4IfTvm.data_max_t))
 {
   if (spi4IfTvm.deskewed)
    {
     trPattern_deskewed();
     spi4IfTvm.training_count = 0;
    }
   if (spi4IfTvm.training_count == spi4IfTvm.data_max_t)
    {
     trPattern_cycles();
     spi4IfTvm.training_count = 0;
    }
 }
 else
 {
  tbvWaitCycle(spi4IfTvm.tdclk, tbvThreadT::ANYEDGE);
 }
 }
 }


 // ***************************************
 // trPattern_deskewed function
 // ***************************************
 void spi4PatternTaskT::trPattern_deskewed()
 {
  //spi4IfTvm.spi4PatternMutex.lock();
  spi4IfTvm.spi4TDataMutex.lock();
```

```cpp
    //
    // wait for Reset to end (if it is asserted)
    //
    if (tbv4StateNotEqual(spi4IfTvm.reset, 1))
      tbvWait( spi4IfTvm.not_reset && spi4IfTvm.anyedge_clk , tbvEventExprT::RWSYNC );


    // Start transaction recording
    spi4IfTvm.spi4IfFiber.beginTransactionH("training pattern");

    spi4IfTvm.flag_deskewed = 1;

//  vector<uint16> data = TrGenArgsP->getDataBy16();
    vector<uint16> data;

    data = spi4IfTvm.old_data;
    //spi4IfTvm.old_data = TrGenArgsP->getDataBy16();


  // Odd Parity Calculation - calculated similarly to Odd Parity Calculation in spi4TDataTask

    unsigned int data_cntr;

    data_cntr = 0;

    data_cntr = (0 << 15) | (spi4IfTvm.curr_eop << 14) | (spi4IfTvm.odd_byte << 13) | (0 << 12) | (0 <<
4) | 0xf;

    spi4IfTvm.dip4 = spi4IfTvm.dip4_calc(spi4IfTvm.invalid, data_cntr, data);

  // Send the control word before the training pattern

    spi4IfTvm.tctl = 1;
    spi4IfTvm.tdat(15,15) = 0;
    spi4IfTvm.tdat(14,14) = (spi4IfTvm.last_data_valid) ? spi4IfTvm.curr_eop : 0;
    spi4IfTvm.tdat(13,13) = (spi4IfTvm.last_data_valid) ? spi4IfTvm.odd_byte : 0;
    spi4IfTvm.curr_eop = 0;
    spi4IfTvm.odd_byte = 0;
    spi4IfTvm.tdat(12,12) = 0;
    spi4IfTvm.tdat(11,4) = 0;
    spi4IfTvm.tdat(3,0) = spi4IfTvm.dip4;

    //tbvOut << " DIP4_PATTERN at : " << spi4IfTvm.tdat(3,0) << tbvGetInt64Time(TBV_NS) << " ns"
<< endl;

    tbvWaitCycle(spi4IfTvm.tdclk, tbvThreadT::ANYEDGE);


    // Send the pattern
    for (unsigned int i = 0; i < 10; i++)
    {
      spi4IfTvm.tctl = 1;
      spi4IfTvm.tdat = 0x0fff;

      for (unsigned int k = 0; k < 10; k++)
        tbvWaitCycle(spi4IfTvm.tdclk, tbvThreadT::ANYEDGE);

      spi4IfTvm.tctl = 0;
      spi4IfTvm.tdat = 0xf000;

      for (unsigned int j = 0; j < 10; j++)
```

```cpp
        tbvWaitCycle(spi4IfTvm.tdclk, tbvThreadT::ANYEDGE);
    }

  spi4IfTvm.tctl = 1;
  spi4IfTvm.tdat = 0x000f;
  spi4IfTvm.last_data_valid = 0;
  spi4IfTvm.invalid = 1;

  spi4IfTvm.flag_deskewed = 0;

  // Record the values in the argument block onto the fiber
  tbvCellT invalid_pattern(1);
  spi4IfTvm.spi4IfFiber.recordAttribute(invalid_pattern, "training pattern");

  // Finish Transaction Recording
  spi4IfTvm.spi4IfFiber.endTransaction();


  //spi4IfTvm.spi4PatternMutex.unlock();
  spi4IfTvm.spi4TDataMutex.unlock();

}



// ************************************
// trPattern_cycles function
// ************************************
void spi4PatternTaskT::trPattern_cycles()
{
  //spi4IfTvm.spi4PatternMutex.lock();
  spi4IfTvm.spi4TDataMutex.lock();

  //
  // wait for Reset to end (if it is asserted)
  //
  if (tbv4StateNotEqual(spi4IfTvm.reset, 1))
    tbvWait( spi4IfTvm.not_reset && spi4IfTvm.anyedge_clk , tbvEventExprT::RWSYNC );


  // Start transaction recording
  spi4IfTvm.spi4IfFiber.beginTransactionH("training pattern");

  spi4IfTvm.flag_cycles = 1;

  vector<uint16> data;

  data = spi4IfTvm.old_data;

  // Odd Parity Calculation - calculated similarly to Odd Parity Calculation in spi4TDataTask

  unsigned int data_cntr;

  data_cntr = 0;

  data_cntr = (0 << 15) | (spi4IfTvm.curr_eop << 14) | (spi4IfTvm.odd_byte << 13) | (0 << 12) | (0 <<
4) | 0xf;

  spi4IfTvm.dip4 = spi4IfTvm.dip4_calc(spi4IfTvm.invalid, data_cntr, data);

  // Send the control word before the training pattern
```

```cpp
    spi4IfTvm.tctl = 1;
    spi4IfTvm.tdat(15,15) = 0;
    spi4IfTvm.tdat(14,14) = (spi4IfTvm.last_data_valid) ? spi4IfTvm.curr_eop : 0;
    spi4IfTvm.tdat(13,13) = (spi4IfTvm.last_data_valid) ? spi4IfTvm.odd_byte : 0;
    spi4IfTvm.curr_eop = 0;
    spi4IfTvm.odd_byte = 0;
    spi4IfTvm.tdat(12,12) = 0;
    spi4IfTvm.tdat(11,4) = 0;
    spi4IfTvm.tdat(3,0) = spi4IfTvm.dip4;

    //tbvOut << " DIP4_PATTERN at : " << spi4IfTvm.tdat(3,0) << tbvGetInt64Time(TBV_NS) << " ns"
<< endl;

    tbvWaitCycle(spi4IfTvm.tdclk, tbvThreadT::ANYEDGE);

  // Send the pattern
  for (unsigned int m2 = 0; m2 < spi4IfTvm.training_m; m2++)
    for (unsigned int i = 0; i < 10; i++)
    {
      spi4IfTvm.tctl = 1;
      spi4IfTvm.tdat = 0x0fff;

      for (unsigned int k = 0; k < 10; k++)
        tbvWaitCycle(spi4IfTvm.tdclk, tbvThreadT::ANYEDGE);

      spi4IfTvm.tctl = 0;
      spi4IfTvm.tdat = 0xf000;

      for (unsigned int j = 0; j < 10; j++)
        tbvWaitCycle(spi4IfTvm.tdclk, tbvThreadT::ANYEDGE);
    }

    spi4IfTvm.tctl = 1;
    spi4IfTvm.tdat = 0x000f;
    spi4IfTvm.last_data_valid = 0;
    spi4IfTvm.invalid = 1;

    spi4IfTvm.flag_cycles = 0;

  // Record the values in the argument block onto the fiber
  tbvCellT invalid_pattern(1);
  spi4IfTvm.spi4IfFiber.recordAttribute(invalid_pattern, "training pattern");

  // Finish Transaction Recording
  spi4IfTvm.spi4IfFiber.endTransaction();


  //spi4IfTvm.spi4PatternMutex.unlock();
  spi4IfTvm.spi4TDataMutex.unlock();

}


// ***************************************
// a task spi4TStat
// ***************************************
spi4TStatTaskT::spi4TStatTaskT(spi4IfTvmT& spi4IfTvm)
  : tbvTaskTypeSafeT<tbvCellT>(&spi4IfTvm, "spi4TStatTask"),
    spi4IfTvm(spi4IfTvm)

{
  // Disable the mutex built into the fiber and turn off automatic
```

```cpp
      // transaction recording
      setSynchronization(FALSE, FALSE);
   }


   spi4TStatTaskT::~spi4TStatTaskT() {}

   //
   // TVM spi4IfTvm : body method for task spi4TStat
   //

   void spi4TStatTaskT::body(tbvCellT *TrGenArgsP)
   {

      //
      // Do one transaction at a time
      //

      spi4IfTvm.spi4TStatMutex.lock();

      //
      // wait for Reset to end (if it is asserted)
      //

      if (tbv4StateNotEqual(spi4IfTvm.reset, 1))
         tbvWait( spi4IfTvm.not_reset && spi4IfTvm.posedge_tsclk , tbvEventExprT::RWSYNC );


      // Start transaction recording
      //    spi4IfTvm.spi4TStatFiber.beginTransactionH("spi4TStatFiber");

      unsigned int temp_status[256];
      unsigned int x0, x1;
      unsigned int dip2_index;
      unsigned int x1_xor;
      unsigned int x0_xor;
      unsigned int stat3Start;

      // Get the value of each of the two Status Bus bits
      x1 = ((spi4IfTvm.tstat(1,1)) !=0)? 1 : 0;
      x0 = ((spi4IfTvm.tstat(0,0)) !=0)? 1 : 0;

      // Wait for the status to become 3
      while (spi4IfTvm.running && ((x1 == 0) || (x0 == 0)) )
      {
         tbvWaitCycle(spi4IfTvm.tsclk, tbvThreadT::POSEDGE);
         x1 = ((spi4IfTvm.tstat(1,1)) !=0)? 1 : 0;
         x0 = ((spi4IfTvm.tstat(0,0)) !=0)? 1 : 0;
      }

      // Remember the time when status 3 began -> to be used for starting training pattern through the
      flag spi4IfTvm.deskewed

      stat3Start = tbvGetInt64Time(TBV_NS);

      // Wait while status remains 3
      while (spi4IfTvm.running && (x1 != 0) && (x0 != 0))
      {
         tbvWaitCycle(spi4IfTvm.tsclk, tbvThreadT::POSEDGE);
         x1 = ((spi4IfTvm.tstat(1,1)) !=0)? 1 : 0;
         x0 = ((spi4IfTvm.tstat(0,0)) !=0)? 1 : 0;
         unsigned int stat3End = tbvGetInt64Time(TBV_NS);
```

```
   if ((stat3End - stat3Start > (spi4IfTvm.cal_len * spi4IfTvm.cal_m * 1)) && ((x1 != 0) && (x0 != 0)))
     {
       stat3Start = tbvGetInt64Time(TBV_NS);
       //tbvOut << " Pattern_task_run at : " << tbvGetInt64Time(TBV_NS) << " ns" << endl;
       //spi4IfTvm.spi4Pattern.spawn();
       spi4IfTvm.deskewed = 1;
     }
 }

  spi4IfTvm.deskewed = 0;


//Calculate DIP2 on the status received
  x1_xor = 0;
  x0_xor = 0;
  dip2_index = 1;

  for (unsigned int k = 0; k < spi4IfTvm.cal_m; k++)
   for (unsigned int i = 0; i < spi4IfTvm.cal_len; i++)
     {
x1 = ((spi4IfTvm.tstat(1,1)) !=0)? 1 : 0;
x0 = ((spi4IfTvm.tstat(0,0)) !=0)? 1 : 0;
       //tbvOut << " x1_and_x0 are  : " << x1 << "  " << x0 << endl;

       x1_xor = x1_xor ^ ((dip2_index == 1) ? x1 : x0);
       x0_xor = x0_xor ^ ((dip2_index == 1) ? x0 : x1);
       dip2_index = 1 - dip2_index;

       temp_status[spi4IfTvm.port_cal[i]] = x1+x1+x0;
       //tbvOut << " TEMP_STAT is  : " << temp_status[spi4IfTvm.port_cal[i]] <<" at : " <<
tbvGetInt64Time(TBV_NS) << " ns"    << endl;
       spi4IfTvm.fl[spi4IfTvm.port_cal[i]] = 1;
       spi4IfTvm.block16_cnt[spi4IfTvm.port_cal[i]] = 0;
       tbvWaitCycle(spi4IfTvm.tsclk, tbvThreadT::POSEDGE);
     }

  x1 = ((spi4IfTvm.tstat(1,1)) !=0)? 1 : 0;
  x0 = ((spi4IfTvm.tstat(0,0)) !=0)? 1 : 0;
  spi4IfTvm.dip2 = x1+x1+x0;  // 0;

  x1_xor = x1_xor ^ 1;
  x0_xor = x0_xor ^ 1;


  unsigned int dip2_calc;
  unsigned int spawn_flag;

  dip2_calc = (dip2_index != 0) ? (x1_xor + x1_xor + x0_xor) : (x0_xor + x0_xor + x1_xor);

  //tbvOut << "DIP2 is : "<< dip2_calc << "  spi4IfTvmDIP2 is : "<< spi4IfTvm.dip2 << " at : " <<
tbvGetInt64Time(TBV_NS) << " ns"  << endl;

  spawn_flag = 0;
// Only if calculated DIP2 matches the received DIP2, status would be updated
  if (dip2_calc == spi4IfTvm.dip2)
    for (unsigned int d = 0; d < spi4IfTvm.cal_len; d++)
      {
// Check if a favourable status change has happened (2->0, 2->1 or 1->0)
      if (spi4IfTvm.status[spi4IfTvm.port_cal[d]] > temp_status[spi4IfTvm.port_cal[d]])
         spawn_flag = 1;
      spi4IfTvm.status[spi4IfTvm.port_cal[d]] = temp_status[spi4IfTvm.port_cal[d]];
```

```cpp
    //tbvOut << "STATUS_stat is : "<< spi4IfTvm.status[spi4IfTvm.port_cal[d]] << "
UPDATED_TEMP is:" << temp_status[spi4IfTvm.port_cal[d]] << " at : " <<
tbvGetInt64Time(TBV_NS) << " ns" << endl;

    }

// If a favourable status update has happened and the Non-Blocking transfer request FIFO is not
empty - try to send the pending data by spawning queueStat task
  if ((spawn_flag == 1) && (spi4IfTvm.portQueue.size() > 0))
    {
      spi4IfTvm.queueStat.spawn();
      //tbvOut << "SPAWN at : " << tbvGetInt64Time(TBV_NS) << " ns"  << endl;
    }


  // Record the values in the argument block onto the fiber
  //    spi4IfTvm.spi4TStatFiber.recordAttribute(*TrGenArgsP, "spi4TStatFiber");

  // Finish Transaction Recording
  //    spi4IfTvm.spi4TStatFiber.endTransaction();


  spi4IfTvm.spi4TStatMutex.unlock();

}


// ****************************************************************
// a task runTStat
// ****************************************************************

//
// Tvm spi4IfTvmT Task runTStat Constructor & Destructor
//
runTStatTaskT::runTStatTaskT ( spi4IfTvmT & spi4IfTvm ) :
  tbvTaskT ( &spi4IfTvm, "runTStatTask" ),

  // TVM reference
  spi4IfTvm ( spi4IfTvm )
{
  // Disable the mutex built into the fiber and turn off automatic
  // transaction recording
  setSynchronization(FALSE, FALSE);
}

runTStatTaskT::~runTStatTaskT () {}
void runTStatTaskT::body ( tbvSmartDataT *TrGenArgsP )
{

while(spi4IfTvm.running)
{
tbvWaitCycle(spi4IfTvm.tsclk, tbvThreadT::POSEDGE);
spi4IfTvm.spi4TStat.run();
}

}


// ****************************************************************
// a task trainingCnt
// ****************************************************************
```

```cpp
//
// Tvm spi4IfTvmT Task trainingCnt Constructor & Destructor
//
trainingCntTaskT::trainingCntTaskT ( spi4IfTvmT & spi4IfTvm ) :
  tbvTaskT ( &spi4IfTvm, "trainingCntTask" ),

  // TVM reference
  spi4IfTvm ( spi4IfTvm )
{
  // Disable the mutex built into the fiber and turn off automatic
  // transaction recording
  setSynchronization(FALSE, FALSE);
}

trainingCntTaskT::~trainingCntTaskT () {}
void trainingCntTaskT::body ( tbvSmartDataT *TrGenArgsP )
{
  spi4IfTvm.training_count = 0;
  while (spi4IfTvm.running)
    {
      tbvWaitCycle(spi4IfTvm.tdclk, tbvThreadT::ANYEDGE);

// This is a free running counter, which is being reset, only when a
// training pattern starts. It is clipped at data_max_t - max number of clock cycles allowed between
two training patterns.

      if ((spi4IfTvm.flag_data == 1) || (spi4IfTvm.flag_deskewed == 1) || (spi4IfTvm.flag_cycles == 1))
        spi4IfTvm.training_count = 0;
      else
        if (spi4IfTvm.training_count != spi4IfTvm.data_max_t)
          spi4IfTvm.training_count += 1;
    }
}

//****************************************
// dip4_calc function
//****************************************
unsigned int spi4IfTvmT::dip4_calc(unsigned int inavlid, unsigned int data_cntr, vector<uint16> &
data)
{

  unsigned int lines;
  lines = data.size();
  unsigned int jj;
  unsigned int t;
  unsigned int dip16;
  unsigned int dip16_1;
  unsigned int dip16_2;
  unsigned int dip16_3;

  dip16 = 0;

  if (invalid)
    dip16 = data_cntr;
  else
  {
  for (unsigned int k = 0; k < 16; k++)
    {
      jj = k;
      t = 0;
      for (unsigned int m = 0; m < (lines); m++)
        {
```

```cpp
        t = t ^ ((data[m] & (0x8000 >> (jj % 16))) ? 1 : 0);
        jj = jj + 1;
      }

    dip16_1 = (((t != 0) ? 0x8000 : 0) >> ((jj)%16));
    dip16_2 = (0x8000 >> ((jj)%16)) & data_cntr;
    dip16_3 = dip16_1 ^ dip16_2;

    dip16 = dip16 | dip16_3;

   }
  }

 return

    (((((dip16 & (0x8000 >> 0)) ? 1 : 0) ^ ((dip16 & (0x8000 >> 4)) ? 1 : 0) ^ ((dip16 & (0x8000 >> 8))
?
1 : 0) ^ ((dip16 & (0x8000 >> 12)) ? 1 : 0)) << 3) |
    (((((dip16 & (0x8000 >> 1)) ? 1 : 0) ^ ((dip16 & (0x8000 >> 5)) ? 1 : 0) ^ ((dip16 & (0x8000 >> 9))
?
1 : 0) ^ ((dip16 & (0x8000 >> 13)) ? 1 : 0)) << 2) |
    (((((dip16 & (0x8000 >> 2)) ? 1 : 0) ^ ((dip16 & (0x8000 >>  6)) ? 1 : 0) ^ ((dip16 & (0x8000 >>
10))
? 1 : 0) ^((dip16 & (0x8000 >> 14)) ? 1 : 0)) << 1) |
    ((((dip16 & (0x8000 >> 3)) ? 1 : 0) ^ ((dip16 & (0x8000 >>  7)) ? 1 : 0) ^ ((dip16 & (0x8000 >>
11))
? 1 : 0) ^((dip16 & (0x8000 >> 15)) ? 1 : 0));

}


//****************************************
// stop function
//****************************************
void spi4IfTvmT::stop()
 {
   while (portQueue.size() > 0)
     {
       tbvWaitCycle(tsclk, tbvThreadT::POSEDGE);
     }
   running = false;
   //tbvOut << " STOP_function at :   "<< tbvGetInt64Time(TBV_NS) << " ns" << endl;
 }



// ***************************************
// spi4IfTvmT methods
// ***************************************
void spi4IfTvmT::create()
{
 //
 // the base constructor automatically record this TVM instance
 // and will delete it at the end of the simulation.
 //
 new spi4IfTvmT();
}

spi4IfTvmT::spi4IfTvmT() : tbvTvmT(),
  tdclk(getFullInterfaceHdlNameP("tdclk")),
  tsclk(getFullInterfaceHdlNameP("tsclk")),
  reset(getFullInterfaceHdlNameP("reset")),
```

```cpp
    tdat(getFullInterfaceHdlNameP("tdat_o")),
    tctl(getFullInterfaceHdlNameP("tctl_o")),
    tstat(getFullInterfaceHdlNameP("tstat")),

    // Initialize the Mutex
    spi4TDataMutex( "spi4TDataMutex" ),
    fullStatMutex( "fullStatMutex" ),
    queueStatMutex( "queueStatMutex" ),
    spi4TStatMutex( "spi4TStatMutex" ),
    portQueueMutex( "portQueueMutex" ),

    spi4IfFiber(this, "spi4IfFiber" ),
    fullStatFiber(this, "fullStatFiber" ),
    queueStatFiber(this, "queueStatFiber" ),

    spi4TData( * this),
    fullStat( * this, spi4TData),
    queueStat( * this, spi4TData),
    spi4Pattern( * this),
    spi4TStat( * this),
    runTStat( * this),
    trainingCnt( * this),
    running(false),
    cal_len(20),
    cal_m(5),
    old_data(),
    //status(),
    MaxBurst1(256,0),
    MaxBurst2(256,0),
    invalid(1),
    last_data_valid(0),
    last_data_bad(0),
    deskewed(1),
    flag_data(0),
    flag_deskewed(0),
    flag_cycles(0),
    data_max_t(65534),
    training_m(1),
    badParity(0),
    fullPort(0),

    // Create the Event Expressions
    anyedge_clk ( tdclk, tbvThreadT::ANYEDGE ),
    posedge_tsclk ( tsclk, tbvThreadT::POSEDGE ),
    not_reset ( reset , 1 )

{}

spi4IfTvmT::~spi4IfTvmT() {}


/**********************************************************************/
```

References

1. "Overview of the Open SystemC Initiative", September 1999, http://www.systemc.org/projects/sitedocs/document/Open_SystemC_datasheet/en/1

2. Karen Bartleson, "A New Standard in System-Level Design", September 1999, http://www.systemc.org/projects/sitedocs/document/Open_SystemC_WP/en/1

3. "User defined coverage - a tool supported methodology for design verification", R. Grinwald, E. Harel, M. Orgad, S. Ur, and A. Ziv. In Proceedings of the 35th Design Automation Conference, pages158–165, June 1998.

4. "Hole Analysis for Functional Coverage data", Oded Lach-ish, Eitan Marcus, Shmuel Ur, Avi Ziv of IBM Research lab, Haifa - http://www.research.ibm.com/pics/verification/ps/holes_dac02.pdf]

5. "The Unified Verification Methodology", White paper at http://www.cadence.com/whitepapers / 4442_Unified_VerificationMethodology_WP1.pdf

6. "Transaction Based Verification", white paper at http://www.cadence.com/whitepapers/transactionbasedver.pdf

7. "Incisive Verification Platform, Using Assertions", White paper at http://www.cadence.com/products/functional_ver/appnotes.aspx

8. "Verification Interfaces and components library reference and user guide", Cadence IUS 5.3 product release.

9. "SCV Randomization", John Rose, Stuart Swan - Cadence designs systems, Inc Oct 2003, http://www.testbuilder.net/reports/scv_randomization.pdf

10. "Synopsys technical bulletin for design and verification", http://www.synopsys.com/news/pubs/veritb/veri_tb.html

11. "A Verification Methodology and environment", http://www.zaiqtech.com/innovation/m_verification.html:

12. "Regression testing during the design process ", John Sanguinetti & David Manley

13. http://www.a-a-p.org

14. http://www.softpanorama.org/Scripting

15. http://www.aptest.com/resources.html

16. http://www.avery-design.com/twwp1.html

17. http://www.verificationguild.com/guild/latest.html

18. http://www.denali.com/products_mmav.html

19. https://www.ememory.com

20. https://www.denali.com/docs/servlet/ememory.drs.EmemoryDmrHome

21. http://www.azanda.com

Books:

1. Professional Verification: The Guide to Advanced Functional Verification in the Nanometer Era by Paul Wilcox.

2. Principles of Functional Verification by Andreas Meyer.

3. Writing Testbenches: Functional Verification of HDL Models by Janick Bergeron.

4. Timing Verification of Application Specific Integrated Circuits (ASICs) by Farzad Nekoogar.

5. System-On-Chip verification, Methodology and Techniques by Prakash Rashinkar, Peter Paterson and Leena Singh, Cadence Design Systems, Inc.2001.

6. System Design with SystemC by Thorsten Grotker(Editor), Stan Liao, Grant Martin, Stuart Swan.

Index

A
ABV xi
adhoc 22
Analog 114
API 179, 182
Applications 86
Architectural checkers 116
ASIC xi, 1, 18, 253
Assertions 109
ASSP xi
ATE 289
ATE Stages 290
B
Backannotating 282
Backpressure 86
Black-Box 240
Block 81
Block TVM 8, 150
Block Verification 8
Buffer Management 87
Bug Rate 79
Bug rate 228
Bug tracking 18, 37, 216
Bugzilla 38
Build Process 18, 35
C
C models 253
C++ xi
Cadence xi, 35
Capacity 85
Central Messaging 82
Checker 314
checker 77
Checkers 8, 76, 81, 150, 161, 193
Chip 81
Code Coverage 18, 41, 236
Code Example 319, 332, 344
Commercial memory models 256

Common Infrastructure 297
Complex Results Checking 101
Concurrency 46, 52
Congestion 86
Constraints 56
Consumer 120
Control path 89
Control path checker 198
Corner Case 90
Coverage 6, 228
coverage 78
Coverage Model 236
Coverage Monitor 240
Coverage Tasks 238
CPU 5, 6, 76
CPU slots 9
CRC 73, 158
Cross Product coverage models 238
CVS 34
Cycle-Accurate Model 98
D
Data structures 25
data structures 161
debug 18
Debussy 36
Denali 39, 252
design cycle 18
Directed random testing 235
document 91
Documentation 18, 26
Doxygen 27
Driver 80, 304
DUT 70, 154
DUV 73, 154
Dynamic Memory Modeling 9, 251
Dynamic Memory modeling 260
Dynamic memory modeling 2
E
ECO 287
EDA 6, 24
EOP 73

Error diagnosis 227
Example 319
Exit criteria 227
F
FIFO 86
Flow 17
format 65
FPGA 18
Free Buffer List 86
FSM 21
Fspecs 68
full timing 277
Functional Coverage 2, 9, 18, 40, 233
functional/cycle accurate 5
Functions 122
FVP 104, 112
G
Gate Level 10
gate level simulation 35
Gate Simulation 275, 277
Generator 75, 82, 299
generator 18
Grey-Box 240
H
Hardware 217
HDLLint 21
Hierarchical Transactions 125
Holes 238
Host Resources 217
HVL xi, 5
HVLs 7
HW/SW 22
I
IC xi
IEEE xii
Incisive 22
inertial delays 284
infrastructure 82, 150, 161
Integration 79
integration 112
Interface Monitors 116

Introduction 7
IP 5
L
Lint 7, 18
Load Sharing Software 218
LSF 40
M
master 8
Master transactors 130
Memory Controller 87
Memory modelling 18
Memory Models 161
Memory models 8, 150
memory models 252
Message Responder 201
Messaging task 161
Methodology 8
Metrics 79
Model 86
Modeling 95
Monitor 310
monitor 18
Monitor transactors 130
Monitoring 228
Monitors 81, 161, 193
monitors 8
Multicast 88
Multicycle path 282
mutex 24
N
ncpulse 284
ncsim 35
ncverilog 35
Negative 86
Nova-Verilint 22
O
Open source 23
Open SystemC 47
Order-Accurate Model 99
OSCI 7, 45
OVA xii

OVL xii
P
Packet 76
packet 73
Parity 73
Performance comparison 273
Performance tests 88
Pipelined 121
PL2 74, 160, 170
PL3 74, 160, 163
Pl3Rx 332
PL3Tx 319
Policing 88
Ports 55
Post Synthesis 275
Post synthesis 2
Post Synthesis/Gate Level Simulation 9
Producer 120
protocol 18
ProVerilog 21
PSL xii
Pull Interface 131
Pulse handling 285
R
Random data generation 46
Random Verification 89
Randomization 56
randomization 48
Rate Monitors 77, 150
Rave 23
Reference models 130
Regression 8, 18, 40, 78, 211
regression 5
Regression Components 216
Regression features 220
Regression Flow 213
Regression Phases 214
Regression Scripts 219
Regression Test 219
Reporting Mechanism 225
response 116

Response checkers 130
Results Checker 98
Revision Control 18
Revision control 32
RTL 6, 96
Run options 229
S
Sampling Event 239
Sanity 83
sc_in 56
SC_METHOD 122
sc_port 56
SC_THREAD 122
Scripting 18, 30
Scripts 78
SCV 2, 7, 25, 45, 78
SCV Randomization 50
SDF 282
self checking randomization 6
Semaphore 24
Shaper 88
Sideband 303
Signal-Level Interface 134
Simple example 299
Simulation 18, 35, 278
simulation 6, 253
Simvisio 37
Slave transactors 130
slaves 8
Snapshot 240
SoC 103
SOP 73
Specman 23
SPI4 74, 160, 338
SpyGlass
    22
SST2 73, 75, 157
Stages for simulation 280
State Machine 141
Statistical 240
Statistics 88

statistics 6
Stim 318
Stimulus 83
stimulus 5, 18, 65, 116
Stimulus File 177, 192
Stimulus generation 281
Stimulus generators 130
Stimulus/Traffic Generation 8, 150
streams 125
Stress 240
SUPERLOG xii
Superlog 23
SureLint 21
SVA xii
synchronization 46
Synopsys xi
System level testplan 70
System Level Verification 156
System Level verification 8
SystemC 7, 8, 23, 64, 95
System-on-Chip xi
SystemVerilog xi, 23
T
tape out 18
Tasks 55
TBV xi, 153
Temporal 240
Test 55, 83, 154
test builder 9
Test Execution 90
Test Plan 7
Test plan 2
Testbench 23, 95
Testbench Components 117
Testbench Structure 54
Testbuilder xi
Testing at ATE 291
Testplan 65
Tests Status 91
tests status 68
thread 24

timing verification 9, 275
TLM xi, 96, 114
TLM. 56
Tools 80
tools 8, 152
Top Level simulation 209
Traffic Generation 170
Transaction 8, 18, 74, 154
transaction 8, 95
Transaction Accurate Model 100
Transaction Level Modeling 96
Transactions 107, 123
Transactor Structure 55
transport 284
Transport delay 285
TVM 26, 74, 80
TVMs 153
U
Unit Delay timing 277
Using SCV 7
UTOPIA 74, 160
UVM 103
V
VCS 35
verification components 22
Verification Metrics 228
Verification Process 7
Verilog 25
Verisity xi
Version control 220
VHDL 25
Video systems 243
W
waveform 18
White-Box 240